Power, Knowledge, and Expertise
in Elizabethan England

The Johns Hopkins University Studies
in Historical and Political Science

122nd Series (2004)

1. John D. Krugler, *English and Catholic: The Lords Baltimore in the Seventeenth Century*

2. Eric H. Ash, *Power, Knowledge, and Expertise in Elizabethan England*

# Power, Knowledge, and Expertise in Elizabethan England

Eric H. Ash

The Johns Hopkins University Press
*Baltimore and London*

© 2004 The Johns Hopkins University Press
All rights reserved. Published 2004
Printed in the United States of America on acid-free paper
9 8 7 6 5 4 3 2 1

The Johns Hopkins University Press
2715 North Charles Street
Baltimore, Maryland 21218-4363
www.press.jhu.edu

An earlier version of portions of chapter 2 appeared as "'A perfect and an absolute work': Expertise, Authority, and the Rebuilding of Dover Harbor, 1579–1583," *Technology and Culture* 41 (2000): 239–68.

Library of Congress Cataloging-in-Publication Data
Ash, Eric H., 1972–
  Power, knowledge, and expertise in Elizabethan England / Eric H. Ash.
      p. cm.
  Includes bibliographical references and index.
  ISBN 0-8018-7992-2 (hardcover : alk. paper)
  1. Great Britain—Politics and government—1558–1603. 2. Power (Social sciences)—Great Britain—History—16th century. 3. Knowledge, Sociology of—History—16th century. 4. Great Britain—Intellectual life—16th century. 5. Expertise—Political aspects—Great Britain. I. Title.
  DA356.A69 2004
  303.48'3'094209031—dc22
                                                    2004003188

A catalog record for this book is available from the British Library.

# Contents

|   |   |   |
|---|---|---|
| | *Acknowledgments* | *vii* |
| | *Note on the Text* | *ix* |
| | Introduction: Expert Mediators and Elizabethan England | 1 |
| 1 | German Miners, English Mistrust, and the Importance of Being "Expert" | 19 |
| 2 | Expert Mediation and the Rebuilding of Dover Harbor | 55 |
| 3 | Early Mathematical Navigation in England | 87 |
| 4 | Secants, Sailors, and Elizabethan Manuals of Navigation | 135 |
| 5 | Francis Bacon and the Expertise of Natural Philosophy | 186 |
| | Conclusion: Power, Authority, and the Expert Mediator | 213 |
| | *Notes* | *217* |
| | *Essay on Sources* | *249* |
| | *Index* | *257* |

# Acknowledgments

The research and writing process for this book has stretched over several years, during which time I have acquired many debts. My archival research in the United Kingdom was generously funded by a grant from the National Science Foundation (SES-9729740), as well as by grants from the Committee for European Studies and the Woodrow Wilson Fellowship Program of Princeton University. Much of the needed revision was undertaken during a postdoctoral fellowship at the Dibner Institute for the History of Science and Technology (2001–2), and I am most grateful for the resources (physical, temporal, collegial, and bibliographic) that were made available to me during my time there. I also wish to thank the Department of History, the College of Liberal Arts, the Office of the Vice President for Research, and the Humanities Center at Wayne State University for their unified support of this project in its last phases.

The staff of Firestone Library at Princeton University, and especially the Interlibrary Services Department, greatly facilitated my research by making such a wealth of research materials locally available to me. The British Library in London afforded me many months of pleasant research, in both its old and its new facilities, and I especially wish to thank the courteous and efficient staff of the manuscript room for all their assistance. Given that I was working at the British Library before, during, and after the complicated process of transferring the library's vast collection from one facility to another, I cannot adequately express my appreciation for their admirable adherence to their moving schedule. The staff at the Public Record Office in London were very helpful in guiding me through their many collections, catalogs, and calendars. I spent a most enjoyable week at the Bodleian Library at Oxford University, and I thank the staff of Duke Humfrey's Library in particular for their patience and kind assistance. In addition to images from the British Library's collections, illustrations in this book come from the Beinecke Rare Book and Manuscript Library at Yale University, the John Carter Brown Library at Brown University, the National Maritime Museum in London, and the Peterborough Cathedral Library (via the Cambridge University Library). I am grateful to the staff at each institution for helping me to track down the images I sought.

I was fortunate during my graduate studies at Princeton to be able to benefit from the mentoring of three faculty advisers, each of whom helped to inspire and shape this project in important ways. Anthony Grafton, my principal adviser, has been ceaselessly supportive and generous with his time and advice. He is responsible for making me think in new ways about patronage and the early modern demand for new intellectual skills, and his detailed comments have helped me to improve every chapter. I am also grateful for his friendship. Peter Lake has taught me much about Tudor governance and deserves special thanks for volunteering his time to initiate me into the mysteries of sixteenth-century paleography. Michael Mahoney has repeatedly shaken me out of my assumptions with his pointed and timely questions, and I have always appreciated his encouragement and his eagerness to talk things over at length. I did not spend a great deal of time in his office while at Princeton, but the time I did spend there was formative.

Several of my colleagues, at Princeton and elsewhere, have read some or all of this book in various drafts and have made innumerable comments that have helped me to correct and improve it; all remaining errors and shortcomings are, of course, my responsibility alone. I am especially grateful to the regular participants in Princeton University's History of Science Program Seminar, including David Attis, Jamie Cohen-Cole, Angela Creager, Gerry Geison, Charles Gillispie, Mary Henninger-Voss, Ann Johnson, Ole Molvig, Suman Seth, and Norton Wise. Nicholas Dew, Robert Goulding, Lauren Kassell, Jill Kraye, Margaret Meserve, and Adam Mosley have all offered me not only their insightful criticism but also their kind hospitality on many occasions when I was overseas. Mario Biagioli served as the outside reader for my doctoral dissertation, and his criticisms and suggestions have helped me turn it into something more. Alison Sandman, Tara Nummedal, and Elizabeth McCahill have all been excellent colleagues and warm friends; I look forward to working with each of them in coming years. I owe Leonard Rosenband, whom I met at the Dibner Institute, a debt of gratitude for his guidance and advice at crucial moments, as well as his friendship in general. I must also thank Robert J. Brugger, Juliana McCarthy, and Melody Herr at the Johns Hopkins University Press for their great patience and constant support in bringing this project to a successful conclusion, Lois Crum for a thorough job of copyediting, and an anonymous reader for some extremely helpful comments and suggestions. Catherine Bogosian kindly helped me with all proofreading.

Finally, none of the educational endeavors that have culminated in this book would have been possible without the love, support, and devotion of my family. I lovingly dedicate this book to my parents, Molly and Earl Ash, and to my grandfather, L. Forrest Hotchkin, a small token of my appreciation for all that they have given me. I regret very much that my father and grandfather did not live to see it published.

# Note on the Text

All dates in this book refer to the Old Style (Julian) calendar, used in England throughout the sixteenth and seventeenth centuries, save that the new year is understood to begin on 1 January, rather than the common Elizabethan date of 25 March.

In all quotations, I have altered the spelling and capitalization to conform to modern American English usage and have expanded all contractions. In a few cases, where the original punctuation was especially obscure or confusing, I have altered it slightly to reflect more modern usage. With respect to titles of works and documents, I have retained the original orthography.

# Power, Knowledge, and Expertise
in Elizabethan England

INTRODUCTION

# Expert Mediators and Elizabethan England

> And whereas in the universities men study only school learnings, in this Academy they shall study matters of action meet for present practice, both of peace and war. And if they will not dispose themselves to letters yet they may learn languages or martial activities for the service of their country. If neither the one nor the other then may they exercise themselves in qualities meet for a gentleman.
>
> HUMPHREY GILBERT, "THE ERECTION OF AN ACHADEMY IN LONDON"

The winter of 1587–88 was an anxious time in England. The realm was preparing for war with the largest and most powerful empire in the world, a political and military colossus controlling half of Europe and spanning most of the New World as well. The ruler of this vast dominion, King Philip II of Spain, was irritated by English involvement in the rebellion of his Dutch subjects, frustrated by repeated English incursions into Iberian trading monopolies, and enraged by English acts of piracy on the open sea and naval warfare on his own coasts. Moreover, Philip wanted badly to strike a blow at Protestant resistance to Catholic domination throughout Europe, or at the very least, to halt the persecution of Catholics in Protestant-controlled areas. By 1587 Spain's relations with England had become so strained that Philip had given up on a diplomatic resolution; the English would have to be beaten in battle, and Queen Elizabeth either forced to capitulate or deposed in favor of a more tractable successor. Pope Sixtus V had been urging such a bold move on Philip for years, believing that Elizabeth was responsible for supporting an alarming number of Protestant uprisings in northern and central Europe.[1]

The Spanish empire was at the height of its power during the 1580s; Spain itself had a population of perhaps 8 million, roughly twice that of England, and its extensive colonial enterprises in South and Central America generated hundreds of tons of silver every year, enough to fund a huge military for campaigns both overseas and closer to home. The Spanish-Habsburg army was the largest, best trained, most experienced, and most feared in all of Europe. The professional, battle-hardened veterans stationed in the Low Countries in particular, under the command of the brilliant Duke of Parma, would certainly prove more than a match for England's small, inexperienced, and poorly armed civilian militia.[2] But to bring these vast resources to bear against the English, Spanish forces would first have to cross England's most imposing line of defense: the natural moat of the English Channel.

For months Philip had prepared for this enormous and complex mobilization, the largest seaborne military operation in history. His assembled naval forces, the "Invincible Armada," were supposed to sail up the English Channel, beat back the English defensive fleet, rendezvous with Parma's army, and convoy them across to England, where they would land in the Thames estuary and march on London at once. Although Philip had hoped the fleet would sail in the late fall or early winter of 1587 and did all that he could to hurry the preparations toward that end, the Armada was delayed throughout the winter by inadequate supplies, insufficient armament, and poor organization. Yet by the spring of 1588, under the guiding organizational hand of the mission's naval commander, the Duke of Medina Sidonia, the Spanish Admiralty had assembled a formidable fleet of 130 sail, including ships from all parts of Spain's maritime empire. The Armada finally weighed anchor from Lisbon on 29 May 1588 and began its journey northward toward the Channel.

Virtually everyone in England, from the queen and her advisers down to the multitudes of common English cloth workers (demand for whose services had plummeted through the disruption of the Spanish cloth trade), knew that a Spanish invasion attempt was imminent. There had already been a few scares and false alarms during the winter, which had revealed many flaws in England's readiness to meet the threat. But as the spring approached, the English seemed as well prepared as they could be. Understanding very well that the sea provided them with their first and best defense, for years the Royal Navy had been preparing for a decisive battle, in which the officers hoped to thwart any invasion force before it could reach its intended landfall. If the martial prowess of England's civilian militia was dubious, by 1588 the might of the English navy was unquestioned. Though slightly smaller than that of its Spanish rival, the English fleet assembled to meet the Armada included some of the newest and finest ships in all Europe. Through the radical administrative reforms of the navy's treasurer and comptroller, John Hawkins, the Royal Navy's maintenance budget had been raised to four thousand pounds per year, and a new frontline warship had been built every year since 1583, while many older royal vessels had been completely refitted or rebuilt.

For both the new and the rebuilt galleons, Hawkins had insisted upon an innovative style of ship design, based on a new philosophy of combat at sea. Previous naval engagements—especially those fought in the Mediterranean, for which most of the Spanish ships were best suited—had featured large vessels carrying small guns and scores of men. The giant ships sailed or rowed into close proximity to one another, used small, short-range cannon and small-arms fire to kill as many men as possible, and then decided the battle through boarding and hand-to-hand combat. Hawkins rejected this style of naval warfare in favor of keeping enemy ships at bay. His newly designed, "race-built" English galleons were comparatively smaller, thinner, and closer to the water, but they were faster and far more maneuverable than larger, more traditional vessels of a similar class. They could sail almost straight into the wind and gain the weather advantage over the enemy at will; in battle they literally sailed circles around the Armada vessels. They also carried fewer crewmen on the whole but more cannon with heavier shot and much longer range, intended to kill ships rather than men. Instead of closing with the enemy, the English race-built galleons were designed to outsail, outmaneuver, and outgun enemy ships, harassing them from a distance and preventing them from ever reaching England's shores. This principle guided the construction and rebuilding of all royal vessels after 1583.

The officers and crews manning the English fleet were also at the top of their art. Although English mariners had lagged behind their Iberian rivals in experience and training between 1500 and 1550, in the succeeding decades they had mastered the nautical advances introduced by the Spanish and Portuguese and were starting to produce some innovations of their own. Navigational instruments such as the cross-staff, the mariner's astrolabe, and the plane chart had become standard equipment for English pilots, and their owners were adept at using them to guide their ships at sea. More importantly for the battle itself, the English had gained a great deal of valuable nautical experience by 1588. English ships had sailed in all regions, climates, and weather conditions and had even circumnavigated the globe twice. Along the way, their crews had fought countless small skirmishes with Spanish merchantmen and warships, learning their naval tactics and capabilities; they knew how to win naval battles with the Spanish and how to avoid those they could not win. There was, indeed, very little one could do with a stout sixteenth-century galleon that English mariners had not already done.[3]

By the spring of 1588, the English Privy Council and the administrators of the Royal Navy had received enough intelligence concerning Philip's invasion plans to put everything in a state of readiness. Every royal ship was freshly repaired, caulked, tarred, and scraped free of barnacles. Every ship was stocked and supplied for a three-month engagement. Every crew stood at only half strength—a shrewd cost-saving measure ordered by the queen—but the crewmen ashore were ready to report for duty at a moment's notice. The ordnance aboard each ship was formidable and primarily intended to damage ships in long-range combat, in accordance with Hawkins's naval philosophy. The majority of English guns were cast from iron in one of several domestic gun foundries that were the envy of Europe; indeed, a notable proportion of Spain's ordnance also consisted of smuggled or captured English guns. Several ships probably carried some superior brass or bronze cannon, also cast in England from copper, zinc, and tin—all of these metals were readily available from English mines, many of which had been newly discovered and exploited within the previous twenty-five years.

The Royal Navy's vessels did not sail into battle alone; just as the queen and her advisers had vigorously supported English overseas trade for decades, now they asked the English mercantile community to contribute to the defense of the realm. The Privy Council saw to it that the royal fleet was heavily supplemented by private merchant ships, many of which were nearly as impressive as the queen's. Over the previous decades, the English merchant marine had often engaged in naval combat to protect or enlarge its trading interests, and several of the scores of merchantmen volunteered or commandeered for war with Spain were designed to maximize speed and maneuverability. Most were also impressively well armed, allowing their crews to defend themselves from pirates, force their way into Iberian trading monopolies, or even capture foreign prizes as pirates and privateers themselves. All of this coincided perfectly with Hawkins's new naval philosophy, and the contribution of the merchantmen to the fleet played a critical role in battling the Armada.

To alert and mobilize the English merchant marine for battle, on 1 April 1588 the Privy Council sent urgent messages to virtually all the port towns in the realm, ordering them to furnish to the war effort whatever ships and men they had available. Some ports were slow to reply and reluctant to help, complaining of unbearable costs or claiming that their best ships were already abroad on trading voyages. The town of Dover, however, answered the call within twenty-four hours. Dover was home to an ancient yet troubled harbor in southeastern England that had been completely redesigned and rebuilt during the 1580s using the Crown's customs revenues; the new jetties of the harbor mouth, in fact, were still under construction in 1588. Mindful, perhaps, of the enormous financial and organizational support his town had so recently received from the queen and her ministers, the master mariner William Courteney of Dover wrote to Francis Walsingham, the principal secretary of the Privy Council, to offer up whatever resources the town and port could muster.

Courteney volunteered the services of up to twenty of Dover's best hoys, small ships that could be manned by crews of roughly twenty-five and were mainly used to transport passengers or small cargoes. Yet Courteney explained in his letter that despite their size, they could render valuable service even against the giant Spanish warships, within the context of Hawkins's battle plans. Like the navy's race-built galleons, the Dover hoys were faster and more maneuverable than larger ships; in a fair wind and with a skilled crew, they could easily fire two or three targeted broadsides in the time it would take a large ship to come about and fire one. Moreover, they could be nimbly handled by a very small crew, leaving more room for guns and gunners to man them, and making them highly economical to keep in an indefinite state of readiness. And because the ships were already conveniently located at Dover Harbor, they enjoyed an ideal vantage from which to respond quickly in reinforcing the main English fleet.[4]

For their part, the Spanish commanders were well aware of the naval strategies and capabilities of the English and the disadvantages from which they themselves would have to fight. Experienced Spanish officers had fought all too many skirmishes and running battles with their English counterparts, and the deadly efficacy of England's ships, mariners, guns, and naval tactics was becoming ever more infamous throughout the Spanish empire. Only the year before, in a bold and successful attempt to delay the Armada's assembly and departure, Francis Drake had dared to sail a small English fleet right into Cadiz Harbor, inflicting heavy damage on the shipping he found there and escaping without suffering any losses himself. The Armada's commanders had every reason to believe that the English would be even more determined and defiant when fighting to protect their own shores. Before the Armada set sail, one unnamed Spanish officer summarized his view of the situation for a papal observer in Lisbon:

> If we can come to close quarters, Spanish valor and Spanish steel (and the great masses of soldiers we shall have on board) will make our victory certain. But unless God helps us by a miracle the English, who have faster and handier ships than ours, and many more long-range guns, and who know their advantage just as well as we do, will never close with us at all, but stand aloof and knock us to pieces with their culverins, without our being able to do them any serious hurt. So . . . we are sailing against England in the confident hope of a miracle.[5]

The outcome of the Spanish Armada's attempt to invade England in the summer of 1588 is, of course, very well known. The Spanish were ill-equipped, poorly supplied, and unprepared for the style of naval combat the English offered them. Their ships could not gain the advantageous weather position over the English, they did not carry enough long-range weapons, powder, or shot, and their crews were weakened and depleted by spoiled provisions. The only advantages they possessed were in their superior numbers of ships and men, but since the English refused to close with them, these advantages could not be brought to bear. The fact that Medina Sidonia had no clear plan to effect a rendezvous with Parma's army in the Low Countries did not help matters. The English commanders were content to stand off and harass the Armada all the way up the Channel, bottle them up in the harbor at Calais, frighten them out of the harbor with fire ships, and then mercilessly scatter them as they fled into the North Sea. Even after they had given up any hope of landing an invasion force and had set a course for home, the Spanish ships were desperately short of supplies, very far from any friendly ports, and at the mercy of the weather in one of the most meteorologically brutal seasons in living memory. Because the ships were forced to chart a torturous course for home that took them all the way around Scotland and Ireland, roughly a third of the fleet never made it back to Spain, while many ships that did return were so badly damaged that they never sailed again.[6]

How much the English fleet actually had to do with defeating the Spanish Armada, and how much the outcome of the battle stemmed from privation, illness, and severe storms, is a matter for debate. My reason for briefly sketching the preparations by the English for invasion is to consider how they managed to mount such an innovative, formidable, and well-prepared defense in the first place. Assembling, supplying, and maintaining a large and viable defensive fleet was an enormous undertaking, requiring a massive, centralized coordination of personnel and material resources, one of which English royal administrators would not have been capable a generation or two earlier.

## Centralization and the Need for Mediators

England's response to the Spanish Armada was the sum of innumerable directly and indirectly contributing parts; no single element was decisive when considered alone, but the combination was impressive and effective. Just as important as the more obvious technical advances in ship design, for example, were a number of background activities and endeavors that took place throughout England over many decades, including the formation of the naval shipyards under Henry VIII; the development of cast-iron gun foundries in the 1540s; the opening of copper mines in Cumberland and zinc mines in Cornwall in the 1560s and 1570s; the reform of naval administration during Elizabeth's reign; the growing maritime experience and mathematical training of English navigators after 1550; the expansion of the English merchant marine; the regular drilling of English naval gunners; and the construction and maintenance of viable harbors for the Royal Navy, at Dover and elsewhere.[7]

One of the most crucial factors in fostering all of England's myriad defensive preparations in the 1580s was the growth of centralized government and administration, especially through the Tudor Privy Council, without which few of the integral contributions to that defense could have been achieved. Within the context of fighting the Armada, England's increasingly powerful and centralized government served to streamline royal naval administration and finance; facilitated radical innovations in ship design and naval tactics; provided the vast increase in customs revenues that paid for the new navy and its harbors; supported the economic activity that encouraged the production or importation of key commodities and the growth of the merchant marine; and in general allowed for the effective coordination of England's modest resources. No large combat fleet could ever have been assembled, supplied, and deployed in so little time by so small a country without such coordination.[8]

The assembly of England's opposition to the Armada is but one example of a much broader trend in the political, economic, and technological development of early modern England, and of western Europe in general. As early modern European regimes grew ever larger, more powerful, and more centralized, the ministers and administrators who ran them sought to control their peoples and territories more thoroughly than they had ever attempted before. Centralization allowed monarchs and their ministers to rule larger areas more actively and effectively, extending royal dominance into the furthest corners of their realms. As royal regimes consolidated their power and authority, they also increased their revenues and bureaucracies and were able to take on challenges that would have been impossible before. Yet a small nucleus of ministers at court could not possibly monitor and manage an entire realm by themselves. The governing of large territories and diverse populations from a central location by a small number of royal administrators necessitated finding a way to extend the royal will from the metropole into the distant, and sometimes resistant, locality. This, in turn, required some delegation of central authority to trusted intermediaries, whose services were vital in ensuring that royal policies and directives were implemented at the local level.

In Elizabethan England, this delegation of power and authority took a variety of forms. Through the administration of royally appointed bishops, for example, the Elizabethan religious settlement was enforced (at least officially) in parishes throughout the realm. Through the lords lieutenant of the local militias and the justices of the peace, both of which offices were appointed by the Privy Council with the authority of the Crown, the traditional power of local magnates was gradually co-opted or supplanted by officers who answered directly to London. By the end of the sixteenth century, Elizabethan privy councillors controlled the selection of several members of Parliament (through their authority both as royal ministers and as powerful local landowners in their own right) and often took a hand in deciding a number of local political offices as well.[9] By these and similar means, the Crown's ministers and advisers gradually amassed an impressive and unprecedented measure of political, judicial, social, and religious control at the local level throughout the whole of England. The point should not be overstated: Elizabethan England was still a long way from becoming a modern, centralized, technocratic nation-state, and the Privy Council's drive toward centralization was not uncontested by local figures, including the still formidable nobility and landowning gentry.[10] Nevertheless, during the sixteenth century, the ministers and advisers who served the Tudor dynasty carved out a new, more active administrative role for themselves and came to believe that the Crown's agents could and should play a central role in conducting local affairs throughout the realm.

The spread of power and authority from the metropole into the traditionally more peripheral areas within western European kingdoms presented early modern administrators with a number of new challenges and opportunities; many of these were technological in nature, including most of the ones that were integral to England's military response to the Armada. From dam construction in Italy to water management in the Low Countries, from canal building across France to the refortification of the English coast, the effort and expense of undertaking such works were enormous, and the potential benefits deriving from them even greater.[11] For the projects to succeed, however, the core strategies of early modern centralized management remained indispensable: to maintain control at the center, administrators still had to find subordinate officers they could trust and rely upon and delegate the actual organization and oversight responsibilities to them. To fill this intermediary role, royal administrators recruited and patronized individuals whom they deemed to be reliable experts in the fields in question; I call these figures *expert mediators*.

The expert mediator in early modern Europe was a knowledge broker and facilitator. He served as the intellectual, social, and managerial bridge between the central administrators who were his patrons on the one hand, and the various and far-flung objects of their control on the other. The expert might or might not be an actual practitioner in his field; this, I will argue, was precisely the period when the very notion of expertise began to shift from an emphasis on practical experience to a more abstract type of understanding. In either case, the expert's principal distinguishing characteristic was his claim to mastery of some rare, valuable, and complicated body of useful knowledge that he could place at the disposal of his patrons. He served both to coordinate and oversee a given project on-site and to explain the project's details and progress to his patrons, mediating between the center and the locality and acting as both a surrogate and an interpreter. Through their expert mediators, administrators at the center could take on and keep control of projects and events outside their traditional sphere of power, while still maintaining the efficiency and control generated by centralized management. Expert mediators were an adaptation, applied to the specific needs of technically complex undertakings, of an effective early modern management tool broadly employed by increasingly powerful, ambitious, and centralized regimes.

In Elizabethan England, not all such projects originated with the queen and her royal ministers directly. As prospective English investors (the powerful merchants and aldermen of London in particular) became wealthier and more adventurous with their money, they pooled their resources into private joint-stock companies organized to pursue expensive but potentially profitable opportunities of all kinds. Joint-stock companies were designed to raise large amounts of capital very quickly, while minimizing the risk to each individual investor; they were usually based upon the grant of a royal patent on a new method of production or a monopoly in exploiting a particular trade.[12] Some of the earliest English examples were incorporated to undertake mining operations, to search for and exploit new markets for English goods, and to place the first English colonies in North America. Many of these companies operated with overt, sustained support from the Crown and the Privy Council and were in fact a central part of England's early imperial aspirations. The line between public and private business was not nearly so clear in Elizabethan England as it is today, and privy councillors and royal courtiers often headed the lists of such companies' most prominent shareholders. Because virtually all of the companies' investors were based in London or the immediate vicinity, each company faced the same management challenges as the centralized royal government and so had a need for expert mediation. Corporate investors thus sought out dependable men of their own, experts who could voyage with the explorers or travel to the provincial mines and look out for the interests of the shareholders back home. The needs of early modern capitalist ventures therefore helped to create an even greater demand for expert mediators, beyond the civil needs of royal administrators.

## Empirical versus Theoretical Expertise

Who were the expert mediators that played such a vital role in the centralized management of technologically complex projects in Elizabethan England? What was the origin and basis for their expertise? How did English patrons identify those prospective clients who possessed the expertise they needed to control? Before the development of such modern phenomena as technical education institutions and a professional engineering community, the notion of what constituted expertise—of who was the most qualified person for the job, and why—could be rather fluid. In sixteenth-century England the meaning of the word *expert* was undergoing a subtle but important transformation. Traditionally, like the classical Latin term *expertus* from which it was derived, it meant to have personal experience of something, to be experienced.[13] For example, a harsh but anonymous critic of Martin Frobisher's command during his voyages of exploration (1576–78) wrote of one of the mission's junior officers, Andrew Dier, that he was "so unexpert of the sea, as he was never further from England than France, and Ireland." Although this author clearly did not think much of Dier as a seaman, his comment referred specifically to Dier's relative lack of *experience* in oceanic navigation; it says nothing directly concerning his knowledge or overall competence as a mariner, within the limited areas where he had sailed before.[14]

Yet the sixteenth-century concept of being expert was gradually changing; it began to encompass not just experience but also *skill*, a more abstract and general term. According to the *Oxford English Dictionary*, *skill* denoted a certain "capability of accomplishing something with precision and certainty; practical knowledge in combination with ability; cleverness, expertness." The term *skill* certainly could and often did imply experience, but the connection was not absolute; one did not necessarily need to have done something over and over again (experience), in order to be able to do it well (skill).

Sixteenth-century usage occasionally employed this ambiguous, and increasingly skill-based, definition of *expert*. For example, in extolling the many practical virtues of mathematics, John Dee wrote that complicated legal affairs required "an expert *Arithmetician*," both to understand the laws themselves and "to decide with equity, the infinite variety of cases, which do, or may happen." Dee was not arguing that English lawyers and judges should be well-practiced mathematicians but rather that they should possess a certain amount of mathematical *skill* to do their jobs most effectively.[15] Similarly, in a document pertaining to the early days of the Company of Mines Royal, the German mine manager Daniel Hechstetter was described as "[a] man very expert with knowledge and understanding of all manner of mines of metals, and minerals."[16] Although Hechstetter did in fact have considerable personal experience in mining and mine management, his expertise was associated in this instance with his "knowledge and understanding"—in other words, his skill. And in praising Stephen Borough's success in exploring the Arctic seas of northern Russia, Richard Eden commended "the expert skillfulness of so excellent a pilot."[17] Since Borough was, at that time, a young ship's master on his first command, and since he was sailing in the area in question for the first time, Eden can only have been lauding his skillfulness as a pilot, not his experience.

As skill came to be an increasingly important part of being considered expert, personal, hands-on experience ceased to be a sufficient basis for expertise, and in some cases it was not even necessary. In a literate age, other avenues were available; one might aspire to become expert at some arts through extensive book learning, for example. Likewise, an unlearned craftsman might be highly experienced in his field but still lack the sort of cleverness and deeper understanding of his craft implied by *skill*, and so not be considered fully expert. As a result, an aspiring client seeking to impress potential patrons might claim to be expert in some art of which he was not even a practitioner himself, solely on the basis of his lofty perspective on it. Emphasizing the value and rarity of true skill as opposed to common experience, he might portray his own knowledge as superior to the vulgar empiricism of mere practitioners, because he could claim to comprehend why and how things worked the way they did—a more fundamental, versatile, and potentially useful form of expertise. This shift in emphasis, from experience to skill, often created social tensions between practitioners and those who sought to rise above them.

This ongoing transformation of the concept of expertise, and hence of the defining characteristics of the expert, resulted at least in part from the early modern demand for expert mediation. The patrons of expert mediators—the royal administrators and corporate investors who relied upon their services—were not explicitly concerned with redefining expertise, of course. They saw their involvement with experts as being purely functional; they needed experts to help them achieve their goals as successfully and economically as possible. Their most important consideration in selecting the individuals to help them was their assessment of the self-styled experts' ability to translate their knowledge and skills into effective action. The patronage networks of Elizabethan London were deeply concerned with practicality and profits, action and results, an emphasis that would seem at first to privilege a more experience-based definition of expertise.[18]

Yet despite their strong interest in practical results, Elizabethan patrons were disinclined to associate with the sort of base, unlearned practitioners who had the best claims to empirical mastery of their crafts. The patrons were, after all, wealthy London merchants, worldly statesmen, and educated courtiers; many had spent time at one of England's universities, where they would have been exposed to the highly literary style of humanist-inspired education prevalent there.[19] One influential trend in the humanist tradition saw education as the key to leading an active, useful life. Such humanists believed that practical knowledge could (and should) be systematically abstracted, separated from its unlearned practitioners, and organized so as to render it fit for the attention of the educated classes; this branch of humanism was especially concerned with artistic, natural philosophical, mathematical, and technological pursuits.[20] Most English patrons' predilection for action and results was therefore tempered with a respect for scholarly learning; they viewed education as an effective means of attaining practical ends. Rather than turning to common practitioners, then, Elizabethan patrons were an enthusiastic audience for the abstracted, text-based, theoretical version of expertise that began to flourish during the sixteenth century in England. The experts they patronized often shared their humanist educational background, with its proclivity for codifying useful knowledge. One successful patronage strategy among aspiring English experts, therefore, was to present themselves as learned men, whose knowledge was more practical and valuable to their patrons because it was founded upon a more fundamental understanding than mere practitioners could claim to possess. Their portrayal of their expertise thus reflected the humanism of the patronage circles they hoped to join.

The contemporary emphasis on scholarly achievement as a means to attain practical ends appears in several humanist works addressing educational reform. Some authors even lauded the manual arts, encouraging their readers to seek knowledge in the workshop as well as the study and to apply their learning to practical endeavors. Education, they argued, could help to perfect the arts, and learned men thus rendered a great service to the commonwealth in applying their intellectual resources to mastering and improving them. Juan Luis Vives, for example, in his educational treatise *De tradendis disciplinis*, suggested that students should devote some of their attention to solving problems in fields such as weaving, navigation, and agriculture. Men of learning, he wrote, "should not be ashamed to enter into shops and factories, and to ask questions from craftsmen, and get to know about the details of their work."[21] Thomas More's book *Utopia* presented the fable of an ideal commonwealth as a satirical commentary upon early-sixteenth-century English society. In the land of Utopia, More wrote, virtually every citizen was responsible for fulfilling a quota of agricultural and craft labor, whatever their educational or social status. Although the most intellectually talented Utopians might be "permanently exempted from work so that they may devote themselves to study," the members of this elite class were promoted from among the ranks of common laborers, and they were quickly returned to their former employment if their scholarly endeavors proved disappointing.[22] And in France, François Rabelais had his fictional title character Gargantua undergo a rigorous (though humorous) "pursuit of humanistic learning and honest knowledge," which on rainy days included regular instruction in useful arts such as goldsmithing, printing, and dyeing.[23] These books were certainly not written to flatter an audience of actual craft practitioners, most of whom could not have read them in any case, but rather for the edification of wealthy, powerful, and educated patrons.

One interesting English proposal for educational reform was written in the mid-1570s by the professional soldier, explorer, and Elizabethan courtier Humphrey Gilbert. The step-brother of Walter Raleigh, Gilbert was in many ways the epitome of the learned practical man. Educated at Eton and Oxford, he had gone on to fight with distinction in campaigns in France, Ireland, and the Netherlands and had served as both a provincial governor in Ireland and a member of Parliament in London. During the 1560s and 1570s, when he was not actively engaged in royal service, Gilbert put his learning to work by drafting "sundry profitable and very commendable exercises" for the queen, hoping to win her support for an expedition to colonize North America and discover a northwest passage to Asia.[24] He actually managed to found a short-lived colony on the eastern coast of what is now Newfoundland, but he perished on his return voyage to Plymouth in 1583; he was last seen in rough seas on the rear deck of his ship, the *Squirrel*, reading a book. In addition to his petitions in support of North American colonization, Gilbert drafted a proposal for a new school, intended to educate young wards of the Crown and other children of the Elizabethan nobility and gentry.[25] In its focus on practical, vernacular education, Gilbert's proposal anticipated the foundation of London's Gresham College by more than two decades.

In his proposal for the new academy, Gilbert explicitly stressed the growing need of the court and the realm for capable young men of action. The traditional universities at Oxford and Cambridge, he argued, focused too extensively on "school learnings," which represented only a fraction of what an educated English gentleman ought to know. At Gilbert's academy, students would also be required to study "matters of action meet for present practice, both of peace and war . . . for the service of their country" (fol. 6 v). The curriculum of instruction Gilbert recommended showed a strong bias toward the active and the practical, as opposed to mere book learning for its own sake. Although the classical languages of Latin, Greek, and Hebrew were to be taught, for example, Gilbert also called for instruction in more immediately useful foreign languages, such as Spanish, French, Italian, and German, subjects not included in a traditional English university education (fol. 5 r). Moreover, students were to hone their oratorical talents not in academic Latin, but in English, in order that their skills would be more directly applicable after their education was completed. "[I]n what language soever learning is attained," Gilbert wrote, "the appliance to use is principally in the vulgar speech, as in preaching, in parliament, in council, in commission, and other offices of [the] Common Weal" (fol. 2 v).[26] It made little sense, therefore, to neglect practical vernacular training in favor of a language that was rarely if ever useful to the educated man of action.

Other subjects were to be similarly practical in their orientation. The academy's two readers in mathematics were supposed to concentrate on the real-world applications of their subject, the one to "embattlings, fortifications, and matters of war, with the practice of artillery," and the other to "the art of navigation, with the knowledge of necessary stars, making use of instruments appertaining to the same" (fols. 3 r–4 r). The readers in law were to focus not just on legal theory, but also on the actual practices involved in pleading cases and on the holding of judicial or political offices such as justice of the peace, "for through the want thereof the best are oftentimes subject to the direction of far their inferiors" (fol. 4 v).[27] The readers in medicine were to teach the mixing of medicines and especially the common practice of surgery, "because, through want of learning therein, we have very few good surgeons, if any at all, by reason that surgery is not now to be learned in any other place than in a barber's shop" (fol. 4 v). Instruction did not end in the classroom, though; outside of their lectures, Gilbert expected his students to be personally exercised and skilled in all manner of wartime activities, from riding and shooting to pitching a camp and firing a cannon (fol. 3 v).

Gilbert also stipulated that the instructors themselves were to benefit the realm, beyond their teaching duties, through the direct and continued application of their skills and knowledge. The readers in law and the arts were to "set forth some new books in print, according to their several professions," every six years. Those with linguistic skills were to "print some translation into the English tongue of some good work" every three years.[28] The natural philosophers had to perform experiments, "to search and try out the secrets of nature . . . to the end that their successors may know both the way of their working and the event thereof, the better to follow the good and avoid the evil" (fol. 4 r). In oration, warfare, navigation, legal practice, medicine, and language study, Gilbert's proposed academy was designed to generate experienced men of action, trained not for the *vita contemplativa* of further scholarly study, but for the humanist-inspired, courtly *vita activa*, through which they might best serve their queen and commonwealth as educated gentlemen. Their education was intended to prepare them broadly for service in learned, gentlemanly, military, and even some artisanal fields of endeavor. Although Gilbert's academy never got beyond the planning stage, Elizabethan patrons continued to value precisely the sort of action-oriented advisers and clients his curriculum was meant to produce.

Despite the interest of Gilbert and other humanist authors in practical and technical subjects, however, humanism remained at root a literary movement; its adherents naturally turned to books as a means of teaching, learning, and improving the arts they studied. By the end of the sixteenth century, humanist-trained authors throughout Europe had written a vast number of texts describing the skills and knowledge involved in myriad arts.[29] Although many were didactic in style, these were not the sort of books that a master craftsman might use to train a young apprentice. Their language and style were meant to appeal to a readership that probably had little opportunity, reason, or desire to put such knowledge to direct use, and the treatises seldom contained sufficiently detailed instruction to serve that purpose in any case. Instead, early modern humanist technical treatises were written to publicize their authors' mastery of the arts they addressed and to alter the very nature of those arts, elevating them above their traditional, vulgar, unlearned craft status to render them apt fields for leisurely study by a cultured, educated audience.[30]

The publication of such treatises illustrates a mounting tension between two different attitudes toward practical knowledge—specifically, whether such knowledge should be kept secret or communicated openly. Pamela Long has shown that in the context of late medieval urbanism, artisans not only grew in number and economic power but also developed a proprietary attitude toward their craft knowledge, a phenomenon manifested through craft secrecy and patents. During the fifteenth and sixteenth centuries, however, social and economic interactions between patrons, humanists, and craftsmen gave rise to an alternate strategy: by revealing some of their craft secrets, those who controlled such knowledge might benefit more from patronage than from the practice of their craft per se. This new way of exploiting craft knowledge lead to the production of humanist-inspired technical treatises, written not for the use of practitioners but to attract the interest of prospective patrons and to demonstrate publicly the authors' mastery of the art in question. Moreover, by appropriating craft practices into their textual, scholarly outlook, the authors of such books laid claim to craft knowledge as a subject worthy of the attention of gentlemen and scholars. They sought not merely to describe but to reconceive craft practices in terms of abstracted theoretical principles, to organize and codify them textually, in order that they might be more effectively mastered, perfected, and controlled. Their works on practical subjects were meant to provide not only instructions for *how to do something* but also (and more importantly) a deeper understanding of *why things work as they do*.[31]

The drive to theorize and codify the practical arts for the benefit and pleasure of educated and cultured patrons helped to define the identity and role of the expert mediator. The learned readers who made up the principal intended audience for the humanist-inspired, abstracted treatments of the practical arts were the patrons who bestowed their financial and social support on those who offered them useful knowledge and skills. These patrons had large territories to rule, corporations to govern, and complex projects to manage, and they needed to find expert mediators to help them. They were themselves the product of a humanist approach to education and respected the ideal of the learned man of action. It is not surprising, therefore, that they often adopted a humanist approach in deciding which type of experts would be of greatest use to them. Mere craftsmen were increasingly seen as unequal to the tasks at hand. Their limited focus on empirical practice was no match for a broader, abstract understanding when it came to managing complex technical projects, and they therefore could not compete for the personal and professional trust of patrons, so vital to fulfilling the mediator's role.

The claimants to theoretical, skill-based expertise, however, could presumably be counted upon to manage all aspects of a project, to improvise and adapt their broad, fundamental knowledge to whatever contingencies they might face, and to direct the efforts of experienced but subordinate craftsmen accordingly. Their role was not to practice the art in question but to *comprehend* it and to make it available and accessible to their patrons, giving them firmer control over it. The higher degree of education and culture usually possessed by this new class of learned experts served as a shared basis for communication with wealthy patrons and powerful administrators, who were thus more comfortable in extending their trust to the experts as qualified and dependable mediators. Those hoping to receive patronage as experts, therefore, were better off in portraying themselves not as experienced practitioners but as masters of the theoretical principles underlying the practitioner's skills. The most successful expert mediators were able to differentiate themselves from common craftsmen and attain a superior status as the select recipients of courtly and corporate patronage.

Expert mediation thus changed the relationship between patrons and craft practitioners, connecting them in some ways while dividing them in others. Experts did play an important role in bringing both groups into more productive contact with one another, bridging the gap between them by being able to communicate effectively with both sides. Yet they accomplished this by inserting themselves squarely between the two, becoming an obligatory layer of expertise through which all communication had to pass. This stratum of expertise was created largely at the expense of the practitioners, who were increasingly recast as mere empiricists, useful for their practical experience but lacking in true skill and higher understanding. Their divergent attitudes toward secrecy and craft knowledge reinforced the separation between traditional craftsmen and the new class of experts. Whereas craftsmen continued to view their knowledge as proprietary and sought to benefit from its application, experts needed to demonstrate publicly their superior comprehension and control of it, placing it conveniently at their patrons' disposal.

## The Negotiation of Expertise

The goal of this book is twofold: first, to examine the growing reliance upon expert mediators as a tool of centralized management on the part of Elizabethan administrators and investors; and second, to trace the evolution of expertise from a purely empirical to a more theoretical foundation. I will argue that those who were most successful in obtaining patronage as expert mediators used the growing demand for their services to increase their own intellectual and social status and that the patronage process played an important role in redefining what it meant to be an expert.

Chapters 1 and 2 look at case studies from the middle of Elizabeth's reign—the copper mining operations opened in northern England by the Company of Mines Royal and the rebuilding of Dover Harbor by the Elizabethan Privy Council. Both cases show the patrons' acute need for reliable technical expertise in undertaking complex projects for the benefit of themselves and the realm and illustrate the administrative dependence upon expert mediators to monitor and control the projects in question. Both cases also demonstrate that merely acquiring expertise was not the same as deploying it effectively. Although procuring the services of the relevant experts might solve a project's most immediate technical problems, successfully *managing* those experts could prove at least as difficult as finding them in the first place.

Chapters 3 and 4 trace the evolution of the basis of expertise from empirical to more theoretical foundations, by looking at the introduction of mathematical technologies to the traditional art of navigation in England. In the half century after 1550, English pilots and ships' masters were exposed to a variety of new, mathematically based instruments and techniques, which shifted the foundation of their art from local, empirical craft knowledge to an abstract set of mathematical principles that could theoretically be applied anywhere. In the process, the mathematicians who created and controlled these technologies greatly enhanced their patronage status by giving lectures and publishing manuals on the newly reformed art. They made their own expertise seem indispensable to their patrons—the wealthy merchants and investors whose financial well-being was linked to successful navigational endeavors. In the meantime, the practicing mariners who actually guided their ships from one port to another lost some of their social and intellectual status, as they were effectively supplanted by the mathematicians, who often portrayed themselves as the only true experts in navigation.

In chapter 5 I offer a reassessment of some of the early natural philosophical works of Francis Bacon (1561–1626), with respect to the "culture of expertise" prevalent at the Elizabethan and Jacobean courts, where Bacon long tried and eventually succeeded in gaining patronage himself. The main tenets of Bacon's natural philosophical reform program were well established by the end of Elizabeth's reign in 1603, long before he composed his more mature and better-known philosophical works, and were shaped by his perceptions of the patronage values of the Elizabethan court. Bacon certainly shared the humanist education and the high regard for the *vita activa* prevalent among his fellow Elizabethan courtiers, and he was fully aware of the importance of technical expertise in the minds of those from whom he sought patronage. He structured his proposals for philosophical reform accordingly, appealing to the patrons' pragmatic, action-oriented outlook and their enthusiasm for the organization and codification of the underlying principles behind practical activities.

CHAPTER ONE

# German Miners, English Mistrust, and the Importance of Being "Expert"

> Wherefore this is to be considered, that if [Johannes Loner] and his company break off, and leave the works, we have no Englishmen that have skill to take them in hand, so that if we will continue them, we shall be forced to seek for other strangers.   UNSIGNED, UNDATED LETTER TO WILLIAM CECIL

> I have heard that Mr. [John] Chaloner is well learned in Georgius Agricola as touching speculation, and therefore may talk or write artificially, but lacking experienced knowledge by daily working, Georgius Agricola is a present medicine to make a heavy purse light.
> WILLIAM HUMFREY TO WILLIAM CECIL, 22 JULY 1565

In the spring of 1566, Thomas Thurland found himself in a bit of a dilemma. Thurland was the English manager of a nascent copper mining operation in the county of Cumberland (now Cumbria), located in the far northwestern corner of England, an office he had held for just over a year. Although the mines themselves were on English soil, and the private, joint-stock Company of Mines Royal organized to exploit them was controlled by English shareholders, the English themselves knew almost nothing about copper mining. And, indeed, hardly anyone else did either, outside the prolific mining districts of German-speaking central Europe; in the sixteenth century, with very few exceptions, the complicated body of knowledge and skills required to find, dig, and smelt copper ore was almost a German monopoly. Thurland and the rest of the English investors in the company understood that if they wanted to pull copper out of the ground in Cumberland, they would have to do it with German mining expertise or not at all.

To work the mines and process the ore, therefore, the English had entered into a partnership with a group of investors in Augsburg—the banking firm of Haug, Langnauer, and Company—who sent scores of their own experienced miners and mine managers to work in England. In return, the Germans were permitted to own one-third of the company's twenty-four shares. Although the skilled German laborers were to be responsible for virtually all of the company's mining operations, the English partners did not fully trust them to work the mines competently and honestly, without direct English supervision. Fearing that they might become utterly disconnected from their own investment, the English stipulated in their original agreement with the Germans in 1564 that English managers of their own choosing should be involved in the works as much as possible. They hoped that by maintaining a constant English presence at the mines, if not actually in them, they could keep a watchful eye from London over the foreign experts commissioned to work them.

This was the reason for Thurland's presence in the village of Keswick, the center of the company's mining operations in Cumberland and a considerable journey from his clerical post as master of the Hospital of the Savoy in London. Exactly why Thurland had been selected as the company's English mine manager in the first place is unclear. He had been educated at Cambridge, where he received a degree in divinity, and had held a series of clerical offices, including an appointment as queen's chaplain to a parish in Salisbury. He had no prior experience with mining or metallurgy, though he had been sporadically involved in negotiations to hire German smelters to refine the debased English coinage in 1560 and had also been active in helping to broker the mining agreement that had given rise to the company.[1] Perhaps, in the absence of any Englishmen with real mining experience, Thurland's personal knowledge of the Germans was enough to qualify him as the best candidate for the job. In any case, Thurland was the English shareholders' choice to safeguard their interests on site, and he had relocated to Cumberland accordingly. Because the German mine managers, Johannes Loner and Daniel Hechstetter, had returned to Augsburg to consult with their own employers, during much of 1566 Thurland was in charge at Keswick all by himself.

Thurland was deeply troubled that spring because he, like the other English shareholders, had his doubts about the integrity of the Germans, yet he had not the faintest idea what he could do about it. As the senior English officer at the mines, Thurland's job was to monitor the progress made by the German miners and report regularly to London; but he did not really know any more about copper mining than the rest of the English did. Although he had been working side by side with the Germans for a year, he had nothing like their considerable practical experience with mining operations and had no way of knowing whether anything they told him was true. In May of 1566, he began to suspect that the Germans had found more in the mines than they were letting on. He wrote of his concerns to Sir William Cecil, the principal secretary of Elizabeth's Privy Council, one of the queen's closest advisers, and a prominent shareholder in the Company of Mines Royal: "I understand, partly by my own conjectures, and partly by other great presumptions, that there be certain ores here, of gold and silver, gotten in the mines which the strangers keep secret to themselves, and are loathe that

I should know of them before [Johannes] Loner's and Daniel [Hechstetter]'s coming."[2]

What was Thurland to do, under the circumstances? He himself lacked the skill and knowledge he needed to assay the Cumberland ore to his own satisfaction, and there simply were no sufficiently expert Englishmen to whom he could appeal. Indeed, had the English possessed such knowledge in the first place, they would not have had to enter into such a potentially disadvantageous partnership. If the Germans *were* lying to him and working the mines dishonestly, they could conceivably hide it from every Englishman in the company, since there was no one expert enough to catch them at it. It was vital, though, that Thurland not offend the Germans without good cause. They had come to England, after all, at the behest of the queen and the English shareholders, when they could easily have found lucrative employment in any of the several well-established copper mines in their homeland. If the Germans were to take offense at Thurland's suspicions, they might simply decide to go home, taking their irreplaceable expertise with them. Handling such a delicate situation required not only a high degree of mining expertise but a great deal of discretion as well, and Thurland knew that unless he received some knowledgeable and timely outside help, he was not equal to the task.

Thomas Thurland's difficulties during the spring of 1566 were not an isolated problem, and his mistrust of the Germans was far from being his own peculiar paranoia. Thurland's letter to Cecil, and indeed his very presence at the Cumberland mines, illustrates the deep and unallayed suspicion that plagued the Company of Mines Royal from the very moment of its formation in 1564. Many of the English partners had serious reservations about investing their fortunes with a group of foreigners, who seemed to spend their money all too quickly and had little enough to show for it in return. Their mistrust was perhaps understandable, given their large financial stakes in the company and the vulnerable position from which they were forced to deal with their German partners, a vulnerability rooted in their general ignorance with respect to copper mining. For their part, the Germans were unwilling to share all of their valuable knowledge with their English partners; they preferred to run the mining works themselves. They were keenly aware of the suspicion with which the English regarded them and resented the implication that they could not be trusted to do their jobs honestly and profitably.

The early efforts of the Company of Mines Royal miscarried in part because of the endemic mistrust and miscommunication between the English partners in London, the German partners in Augsburg, and the miners and mine managers at Keswick. The English shareholders needed German mining expertise to make the venture feasible, and the German miners employed by the company acquitted themselves admirably, producing many tons of copper metal from the late 1560s onward. Yet the presence of the very experts they needed to undertake the mining works created a serious crisis of management for the English, who had no effective means of controlling the expertise they had acquired. Knowing next to nothing about copper production or large-scale mine management, the English partners could not properly monitor or evaluate the Germans' activities and thus had no way of determining for themselves whether their enormous investment was being used to good effect.

The plight of Thomas Thurland at Keswick illustrates the growing need on the part of administrators and investors in sixteenth-century England not only for technical expertise but also for a reliable means of evaluating and controlling it. The English mine managers were intended to act as the shareholders' eyes and ears, protecting their interests where they were incapable of doing so themselves. In short, Thurland and his fellows were supposed to serve as expert mediators between the partners in London and the Germans at the mines. Their job was to place the Germans' proprietary mining knowledge under the control of the investors who paid for it, making sure that all was running smoothly and keeping the company's nervous investors fully informed. Their mediation failed, however, because despite their proximity in observing the German miners, the English managers never succeeded in making the Germans' mining expertise their own. Not only did they personally have trouble overseeing and trusting those they were supposed to monitor; they also found it very difficult to reassure their colleagues in London because they themselves were obviously not experts and so lacked the authority conferred by the perception of expertise.

## The Art and Science of Sixteenth-Century Copper Mining

Sixteenth-century Englishmen were certainly no strangers to the notion of digging metals out of the ground. Tin had been mined in England as early as 3500 BCE and was being exported from Cornwall to the rest of Europe by 2000 BCE.[3] During the Roman occupation of the island, Britain produced large quantities of lead and iron as well as tin, and there is even evidence that the Romans sank mining shafts to search for copper ore in the Lake District, not far from where the early modern mines were located.[4] English mining declined with the departure of the Romans, as it did throughout much of western Europe. During the Middle Ages, the English maintained a limited production of tin, lead, and iron but did not make any serious effort to mine other metals. As a result, by 1560 they had virtually no experience with the comparatively difficult and complicated processes required to mine and smelt copper ores.[5]

Across the Channel, meanwhile, miners in the German-speaking regions of central Europe had managed to open a number profitable copper mines during the Middle Ages; by 1300 there were thriving copper mining communities in Leipzig, Freiburg, and Nuremburg, to name but a few.[6] The German miners did not merely preserve or resurrect ancient technologies but introduced a number of innovations that allowed them to surpass the achievements of their Roman predecessors, including the use of water-, wind-, and animal-powered machinery and several new smelting techniques.[7] The Germans were also pioneers in writing books about mining and the mineralogical arts; of the major mining treatises published during the sixteenth century, all but one had German authors.[8]

The most important single work on mining operations published during the early modern period was the masterpiece of Georgius Agricola (Georg Bauer), *De re metallica*, first printed in Basel in 1556; it had four Latin editions over the next century, as well as three German and one Italian.[9] Although not a miner himself, Agricola served as a physician and burgomaster in various mining towns in Bohemia and Saxony and published a number of original treatises on mining, surveying, assaying, and mining law. In addition to relying upon his own extensive firsthand knowledge, as a humanist-educated author, Agricola also drew liberally from classical sources in his works.[10] His *De re metallica* may not have been the earliest treatise on mining and metallurgy, but it was easily the most scholarly, comprehensive, and influential of its day. The lengthy and expensive book was also lavishly illustrated and included dozens of now-famous woodcuts depicting all manner of tools and machinery used in locating, mining, and processing ore. When considered together, *De re metallica* and the rest of Agricola's works amply demonstrate the vast scope and primacy of German mining and mineralogical expertise during the sixteenth century.

With respect to copper production in particular, early modern German miners surpassed their English counterparts in four main areas, giving them an unmatched degree of expertise that rendered their participation in any English copper mining venture a patent necessity. Their first advantage was in prospecting for copper ores—in knowing what clues to look for above ground in order to find the metals hidden below. In book 2 (of twelve) of his *De re metallica*, Agricola described how various natural phenomena might accidentally expose metallic ores: floods, earthquakes, and even uprooted trees might all remove enough topsoil to expose a hidden ore vein, whereas forest fires could sometimes produce enough heat to melt the metal in a shallow vein, causing it to flow along the ground in tiny rivulets. In addition, certain variations in soil coloration, foliage patterns, and frost damage could indicate the presence of ores close to the surface. An experienced prospector could recognize such subtle signs, which would be lost on the uninitiated observer.[11]

Recognizing the presence of ores in the area was often only the first step in locating the actual veins, however; rivers could carry ore samples miles from their source, for example, and a prospector had to be able to estimate how far such samples might have been displaced. Once the prospector thought that he had located a vein of ore through one or more naturally occurring signs, he might also use artificial means to confirm his suspicions—by building a dam across a nearby creek and using the pent-up water to wash away the top soil and expose a vein, for example.[12] Whatever their preferred techniques, German miners had far more experience in prospecting for copper ores than any Englishman could claim. Although the English already knew the location of a few native copper veins before the arrival of the Germans, systematic copper mining on the scale undertaken after 1560 in England would have been impossible without the efforts of experienced German prospectors.[13]

The second advantage of the Germans was in the technology of extracting ore from the ground once it had been located. Although dug entirely by hand, German mines were exceptionally deep, and they often used complicated networks of shafts, tunnels, and adits to provide access to ore veins, create ventilation, and allow for the drainage of water. Far from merely recovering the labor-intensive methods used by the ancient Romans, the German miners of the sixteenth century introduced elaborate machines, including multistage pumps to drain flooded mine shafts, bellows and other means of creating artificial ventilation, and complex lifting devices to bring the ore to the surface, all of which were powered by draft animals, windmills, or running water.[14] Book 6 of *De re metallica*, dedicated to the digging and recovery of ore, is filled with detailed descriptions and illustrations of all manner of ingenious machinery. The mechanical principles employed in the devices were ancient, but their application in mines was an early modern development, allowing German miners to dig, drain, and ventilate deeper mine shafts than had ever been possible before the sixteenth century, and certainly far deeper than any English mines of the time.[15] Digging, maintaining, and working such mines required an enormous amount of mechanical and mining expertise, both theoretical and practical. By 1560 the Germans had been working deep mines for generations; the English simply had no comparable experience.

After the ore was located and excavated, it had to be processed; smelting tech-

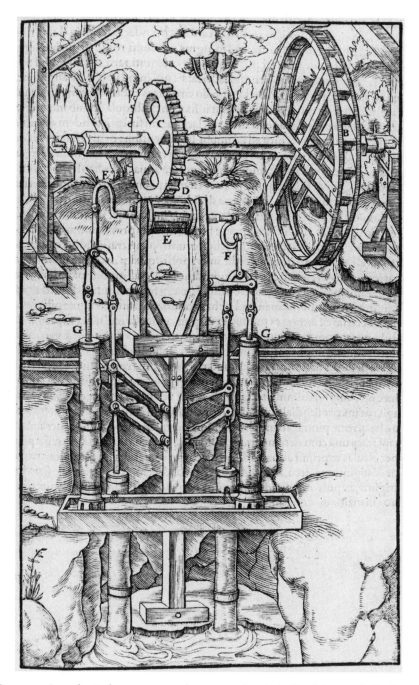

*Figure 1.1.* A mechanical, water-powered pump, used to drain flooded mine shafts. Some shafts, like this one, were too deep to be cleared by a single pump and therefore required a series of pumps and troughs to be placed at regular intervals, one for roughly every twenty-nine feet of elevation. Georgius Agricola, *De re metallica* (Basil: 1556), 147. Photo courtesy of the Beinecke Rare Book and Manuscript Library.

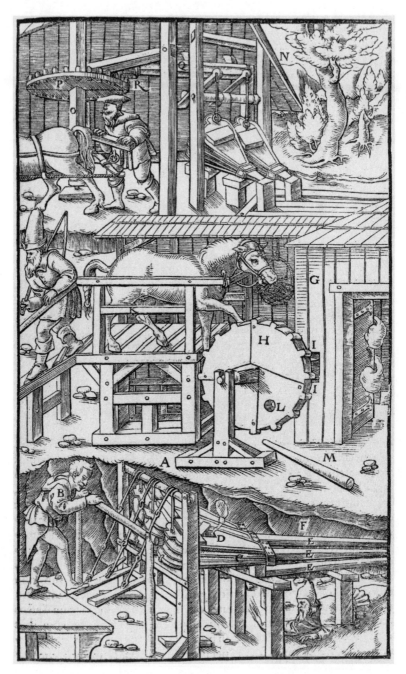

*Figure 1.2.* A series of ventilation machines, powered by men and animals. The complicated gearing mechanisms used motive power to work large bellows, blowing fresh air into the mine shafts. Georgius Agricola, *De re metallica* (Basil: 1556), 169. Photo courtesy of the Beinecke Rare Book and Manuscript Library.

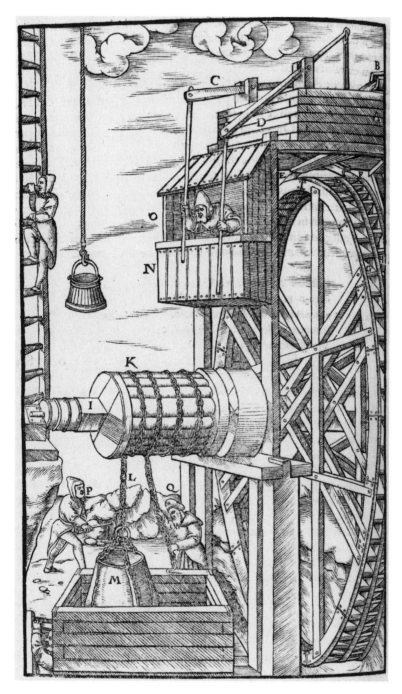

*Figure 1.3.* A water-powered lift, used to bring dirt, rocks, and ore to the surface. The man in the booth controlled the direction of water flow over the paddle wheel and thus the raising and lowering of the large bucket. Georgius Agricola, *De re metallica* (Basil: 1556), 158. Photo courtesy of the Beinecke Rare Book and Manuscript Library.

niques were the third advantage the Germans held. Although the most basic methods of smelting copper had been known throughout Europe since the dawn of the Bronze Age, sixteenth-century German innovations had made the process a great deal more efficient and increased both the quantity and quality of the refined copper. By midcentury, German smelters had developed animal- and water-powered stamp-mills that ground the ore into much smaller pieces than was possible by hand, vastly improving smelting efficiency. In addition, they developed a series of increasingly fine sifting devices (screens, jigs, and washes) in order to obtain the smallest and most uniform ore fragments, sending the larger pieces back for further stamping. Once the ore had been stamped and sifted, it was roasted one or more times prior to the actual smelting. Roasting differed from smelting in that the fire never got hot enough to melt the metal; it was a preliminary step, the limited intent of which was to remove some of the ore's contaminants, another sixteenth-century innovation.[16]

After sufficient stamping and roasting, the ore was finally smelted in furnaces; the metal was at last heated to its melting point, separating it from the remaining impurities of the ore. Early modern German smelters added bellows to their furnaces, which gave a forced air blast and allowed for the construction of still larger furnaces, burning at higher temperatures.[17] Once the copper metal had been separated from the dross, the smelter refined it further by removing other metals from it, especially silver, using a series of specialized chemical reactions and fluxes—the use of lead to extract silver from molten copper was yet another sixteenth-century innovation.[18] So difficult and important was the careful processing of the raw ore that Agricola dedicated nearly half of *De re metallica* (books 7–11) to the preparation, roasting, smelting, and refining of ores and metals.[19] While English miners were certainly adept at smelting the metals they knew well, especially tin and lead, they had virtually no experience in handling copper ore, which is among the most difficult ores to smelt because of the high temperatures required and the variety of metallic and nonmetallic impurities that must be removed.

The fourth advantage that the Germans enjoyed dealt not with the ore and metal themselves, but rather with the administration and management of large-scale, expensive mining operations. Mining, especially in the early stages of opening new mine shafts, was a highly capital-intensive process. Investors had to spend enormous amounts on supplies and labor, with no hope of any return for months or even years, until the mining and smelting works were up and running. Supporting such a venture required not only a huge outlay of capital but also the administrative talent and courage to see it through to profitability. Moreover, once the mines at last became profitable, a number of tricky legal questions were bound

*Figure 1.4.* A mining surveyor, calculating the length and depth of mine shafts. The process relied upon the use of similar triangles, where one length of the larger triangle (HG) and all angles were known. Georgius Agricola, *De re metallica* (Basil: 1556), 90. Photo courtesy of the Beinecke Rare Book and Manuscript Library.

to arise: What if a vein of ore ran beyond the boundary of one's stipulated mining rights? Who could claim the ore? How would the miners, in their shafts deep underground, know when they had crossed the surveyed boundary?

Over the course of many decades, German investors and mine managers had developed several tools and techniques to improve the efficiency and economy of their mining interests. Although they were not the first to experiment with joint-stock companies as a means to raise capital for expensive ventures, German banking firms had been the first to apply the principle to fund mining operations. As with similar companies organized for trade, the joint-stock system of finance limited each partner's individual outlay and risk, while still raising the vast sums necessary to begin the labor-intensive processes of digging mine shafts and erecting smelting furnaces. German surveyors had also learned to adapt the various geometric and trigonometric techniques of their art in order to chart the layout of underground mine tunnels, allowing investors and adjoining property owners to know with certainty the veins to which they could lay claim. Finally, German jurists had gradually codified various practical customs into a body of mining law, which made investment in mining operations a less risky, and hence more attractive, prospect.

The Germans were therefore not only far more experienced than the English in locating, mining, and smelting copper ore; they were also more sophisticated in managing large-scale mining operations and making them profitable.[20] These financial, administrative, legal, and mathematical developments were vital to the successful exploitation of any extensive mine, and they were especially important in starting up a new mine, where property disputes were most likely to arise, financial resources were stretched as far as possible, and any profits were still in the distant future. Agricola, who knew at least as much about managing a mine as he did about digging in one, dedicated much of the first five books of *De re metallica* to managerial issues.[21] The extensive mining "infrastructure" introduced to England by the German investors and mine managers was at least as important to the undertaking as the technical expertise of the German miners. The English possessed nothing so highly developed and relied every bit as much upon sophisticated German management and organization as they did upon German mining and smelting know-how. Indeed, the inability of the English partners to monitor and manage activity at the mines for themselves was the central reason for their mistrust of the Germans and hence for the miscarriage of the company.

## Ignorance, Mistrust, and the Company of Mines Royal

Given that the Cumberland mining operations were to be entirely run by German prospectors, miners, assayers, and smelters and overseen by experienced German mine managers, the English partners really had very little to contribute to the venture other than their money. The Germans, in fact, had apparently not intended for the English to participate even as financiers. Johannes Loner, one of the German mine managers sent to England from Augsburg by Haug, Langnauer, and Company, had also been one of the agents involved in the original negotiations with Queen Elizabeth for mining rights. He later complained that when he and his associates had first suggested the idea of mining copper in England in 1561, they had not sought to take on English partners.[22] Perhaps they did not expect that English investors would wish to involve themselves in such a risky venture, knowing so little about it and never having undertaken any similar projects. In contrast, many German banking and mercantile firms had considerable experience with the technical, financial, and administrative requirements for running profitable mines. Haug, Langnauer, and Company already controlled several copper and silver mines in the Tirol region of central Europe, which they had acquired from the famous Fugger family, giving them ready access to a large and experienced pool of mining labor. The prospective German investors thus believed themselves to be well prepared and amply equipped to take on a new mining interest, whose only novelty for them was its location across the English Channel, without further financial assistance.

Yet the English refused to be left out of the bargain. According to Loner's account, while the Germans were still negotiating with the queen, several would-be English investors demanded to be admitted to the company, "uncalled, and much against our wills."[23] Many of the importunate Englishmen of whom Loner complained were politically powerful; some were members of the queen's Privy Council and enjoyed direct access to Elizabeth herself. Such figures would have exercised considerable control over the all-important grant of royal mining rights, without which the works could not be undertaken. The English investors seem to have used this leverage to force their way into the venture, even obtaining a royally mandated two-thirds controlling interest in the new company. But despite their political clout, the English investors were acutely aware of the serious disadvantage under which they were forced to deal with the German partners—they were always vulnerable to being duped or swindled because of their own near total ignorance of the mining and smelting works. The royally imposed "partnership" between the English and German investors in the Company of Mines Royal was therefore a troubled one from the very beginning.

The English partners soon became alarmed when the two original recipients of the queen's 1563 grant of mining rights, the Englishman Thomas Thurland and the German Sebastian Speydell, petitioned Elizabeth in the autumn of 1564 for an important alteration to their grant. Thurland and Speydell, as the two partners specifically named in Elizabeth's grant, occupied critical positions within the nascent company. Once they had organized the company's affairs, enrolled the other shareholders, collected their initial investments, and begun construction of the mining works, the two would also be expected to manage the mines, maintain accurate financial records, and make sure that the other partners were kept informed of their progress at all times. Yet in their petition to the queen, Thurland and Speydell asked for permission to transfer their full rights and responsibilities to a single German mine manager, Daniel Hechstetter, whom they described as "a man expert in the knowledge and understanding" of copper mining.[24]

Hechstetter was a formidable choice and seemed well suited for such a role. While the Hechstetters had once ranked among the wealthiest and most prominent Augsburg mercantile families, they had fallen into decline by midcentury; and as a second son, Daniel was apprenticed by his relatives around 1540 to learn the mining business. By 1564 he had over two decades of experience in managing copper and silver mines in the Tirol region and was then working for Haug, Langnauer, and Company. He was one of the German assayers sent to England in the previous year to prospect for copper ores and thus had a very good sense of where the ore was located and what would need to be done to extract and smelt it.[25] He appeared amply qualified to oversee the English mining operations, and Thurland and Speydell probably felt that they were placing the company's interests in capable hands.

The English shareholders, however, were uneasy with Hechstetter's appointment. Already aware of their relative weakness in the partnership, they feared that if so much power was concentrated in a single German officer, their entire role in the company would be reduced to practically nothing—simply handing their money over to foreign bankers and laborers, while hoping vaguely for profits to appear at some undetermined time. They therefore submitted their own petition to the queen, in which they made several demands. First, and most immediately, they urged that the new royal grant be amended to include an English officer, to be selected by the English partners, who would share "like authority" with Hechstetter.[26] They simply did not trust a single German mine manager, handpicked and employed by the German partners, to be fair and unbiased in guarding the interests of all of the company's investors. Having an Englishman they could trust on site at all times, the English partners reasoned, would prevent their own interests from being ignored or abused.

The Englishmen's second concern was to overcome their own mining ignorance, the source of their disadvantage within the company. They demanded that "every partner . . . may freely, and thoroughly, at their pleasure, see and understand all the working and manuring [managing] of the said metals and minerals," so that "thereby the truth in their doings may and will be perceived, and their cunning perfectly discovered and learned."[27] So long as the Germans alone possessed all of the skills and knowledge required to run the mines, the English investors would remain peripheral to the affairs of the company. The mines might prosper without English finances, but without German expertise the whole undertaking would necessarily fail. The German partners therefore had power in the company out of all proportion to their numbers, an imbalance that the English feared and sought to nullify by gaining full access to their exclusive expertise.

Nearly all the English shareholders were wealthy London merchants or prominent members of the royal court and the Privy Council; few if any of them ever dreamed of traveling to the far northwestern corner of England to study in person the mining and smelting techniques of the skilled German laborers working there. In subsequent years, very few of them ever made the long and difficult journey even to view and inspect the works. Moreover, their naïveté in believing that mere observation would somehow impart real expertise to the observer was personified in later years by Thomas Thurland's relative helplessness as a mine manager, despite being a longtime overseer at the mines. Nevertheless, the partners' petition of 1564 made clear their single greatest concern: they were forced to bargain with the Germans from a position of weakness because of their own lack of expertise, and it made them uneasy. Their anxiety and suspicion were further illustrated by another demand in the petition, that there must be no secret communication or negotiation between the queen, her advisers, and the Germans. The English partners expected to be kept informed at all times of "all and every authority, license, liberty, and all profits and commodities" associated with the mines.[28]

The English partners received nearly all that they had asked from the queen. On 10 October 1564, Elizabeth issued both new letters patent and an indenture granting royal mining rights jointly to Thomas Thurland and Daniel Hechstetter for twenty years; the grant included the exclusive right to mine gold, silver, copper, and mercury in certain specified English counties and in Wales. Importantly, the two corecipients were to enjoy their patent only so long as they worked together; if they split up, neither was entitled to the patent's protection. The queen, in return for the grant, was to receive a share of all metals produced and to have the first option of purchasing the remaining stock. Thurland and Hechstetter were empowered to form a corporation, hire laborers, acquire land, and purchase supplies at fair market value. They were permitted to import any materials and tools that could not be conveniently found in England and to bring over foreign laborers as necessary to further the works, a point integral to the success of the mines. The foreigners were to have the option of becoming free denizens of England, and the lord keeper was instructed to give them letters patent to that effect, so long as they remained employed at the mines. Finally, in order to raise capital, Thurland and Hechstetter were entitled to enlist twenty-four partners in their venture, at least sixteen of whom had to be English subjects. The English partners were to have a lifetime exemption from all parliamentary taxes, in return for undertaking a financial risk for the good of the realm.[29]

Despite its remote location, Cumberland was probably the most obvious site for the company's prospectors to begin searching for copper—it had long been recognized as a productive mining region and was mentioned as such in various royal grants from 1222 and 1475. Daniel Hechstetter had already assayed ore samples there in 1563, and the strength of his results had secured the interest of his employers in Augsburg.[30] On 26 May 1565, seven months after the company had received its royal grant of monopoly mining rights, Thurland and Hechstetter wrote to Cecil in London to inform him that small-scale preliminary mining had begun and that they had already found large quantities of "copper ore, containing silver." All they lacked to open the new mines in earnest were "skillful and expert arts men" and a sufficient supply of wood for timber and fuel; they therefore asked for permission to bring German miners and smelters to England.[31] Their request was quickly granted, and on 8 July 1565 the Privy Council sent word to the mayor of Newcastle that forty or fifty German miners were on their way there. The strangers were to be well treated and guided to the company's headquarters at Keswick, where Thomas Thurland had already taken up residence the previous March.[32]

Even before the mining works were truly under way, however, a dispute involving metallurgical expertise shook the company. Early in 1565 Thomas Thurland had sent a sample of the Cumberland ore to Cecil, who passed it along to William Humfrey, a shareholder in the company and the assay master of the Royal Mint in the Tower of London, for his opinion as to its value. Humfrey's job at the mint was to assay coins and precious metals to determine their purity, an especially sensitive issue in the mid-sixteenth century, as the debasement of English silver coinage had contributed to the decline of English trade in the Low Countries around 1550. Reestablishing the value of English coins had been a paramount goal of both English merchants and the Privy Council after Elizabeth's succession—indeed, the first negotiations with the Germans over copper mining in England had stemmed from a royal contract of 1560 with Johannes Loner and other German agents to refine English silver coins. Although little is known about his prior training or background, Humfrey's possession of such a vital office at the mint would seem to mark him as one of the most expert Englishman of his day in questions of metallurgy. But the very fact that the Privy Council had had to contract with German bankers in 1560 to get their silver coinage refined speaks poorly of English metallurgical abilities in general.[33] Moreover, testing the purity of refined metal coins was a simple matter when compared with assaying raw copper ore, a much more involved and difficult process requiring considerable care and experience to perform correctly.

Humfrey assayed the ore that Cecil had sent him and reported that it contained at most twenty ounces of copper per hundredweight, a trivial amount of metal.[34] If this result was correct, the Cumberland mines could never recover the costs expended to work them, and the entire venture would be an enormous waste of time, labor, and money. Humfrey's assay, however, was at odds with the much more optimistic one made by Daniel Hechstetter in 1563, on the basis of which he had convinced the German partners to invest in the project. When approached by Cecil with Humfrey's assay results, therefore, Thomas Thurland rejected them outright, declaring flatly that "every hundredweight of that copper ore I sent you holdeth more pounds [of copper] . . . than they [Humfrey] have written ounces."[35]

How could Thurland be so confident? He had even less experience with assaying copper ores than Humfrey and had no hard information on which to base his argument to Cecil except the assurances of Hechstetter and the other German assayers. Meanwhile, the Germans had yet to produce any pure metal (beyond the samples of their initial assays), and they were finding the English ore harder to smelt than the ore they were used to on the Continent.[36] In mid-1565, however, perhaps as a result of living and working so closely with them at Keswick, Thomas Thurland still had complete faith in the competence and integrity of the German miners. He called them "wise men, true dealing men, and artificial men," and believed that they intended "by their industry and great charges to increase this realm with great profit, which hath been long hid." He also pointed out, quite logically, that if the Germans were not firmly convinced of the high quality and value of the English ore, they would hardly have invested so much of their own money in the Cumberland mining works.[37]

Hechstetter, in fact, was correct in his assay, and Humfrey clearly mistaken;

though difficult to work, the ore taken from the Cumberland copper mines was definitely rich in metal. More interesting than Humfrey's error, though, was his response to Thurland's challenge, which reveals much about his confidence in his own abilities as a mineralogical expert. In his letter to Cecil addressing the dispute, Humfrey simply admitted his error and acknowledged his own lack of skill in assaying raw ores, reminding Cecil that "I did also certify your honor how small credit was to be given to the yield [of my assay], in example of the firm ore." He also echoed Thurland in stating that if the Germans were satisfied with the ore's value and believed the mines would be profitable, he would not doubt their conclusions.[38] Given his position at the Royal Mint, Humfrey's failure with the Keswick ore and his ready acknowledgment thereof underscore the deficiency of expertise from which the English had to approach their partnership with the Germans.

Yet Humfrey soon found other reasons not to trust the Germans, having to do not with the mining or smelting of copper, but with the marketing of it. Disposing of copper in sixteenth-century England was not altogether easy, as the demand for it turned out to be somewhat limited. Copper was not used in coinage, except as a means to debase more precious metals; England did not introduce copper coins until the last quarter of the seventeenth century. The Royal Navy's need for cannon was one source of demand, but it did not constitute a dependable, sustained market. Brass and bronze guns were durable and thus did not need frequent replacement, and damaged ones could often be saved and recast; besides, at that time English ships were beginning to capture more guns in battle than they lost.

The biggest and most stable market for copper was not royal but public—though of course the general populace had no use for copper ingots as such. Throughout early modern Europe, virtually all nonmilitary copper consumption was centered on manufactured goods: brass wire, used in making wool cards, and hammered vessels such as pots and kettles. Only the value added by skilled manufacturing labor could allow the metal's price at market to surpass its cost of production, making it a truly profitable commodity.[39] The problem for the Company of Mines Royal was that, as with the mining and smelting of copper ore, the Germans also had a near monopoly on the manufacture of marketable copper goods, including the arts of wire pulling and copper battery (beating the copper into sheets and shaping it into pots and kettles). Moreover, before the 1560s German-controlled mines in central Europe were the only known source of zinc, the metal mixed with copper to make brass. Brass was far more marketable than pure copper, because it was a harder and more durable metal, better suited to the production of cannon and wire especially.

Humfrey determined that for the English copper industry to succeed and prosper, it would have to tap into the lucrative market for manufactured copper goods. Either the English would have to learn the arts of wire pulling and copper battery themselves, or else these tasks would have to be undertaken by still more Germans, to be employed by the company. He also realized that if finished copper products could be produced near an English port, the enormous costs of overland transportation could be avoided when selling them in Continental markets. This would be no small advantage, since a major component of the retail price of such goods in the 1560s was the high cost of transporting them from mountainous central Europe to a port/market city like Antwerp. Humfrey believed that the manufacture of copper products in Cumberland would allow the English to undercut the contemporary German prices on such goods by as much as half, providing enormous profits for the company. He also hoped that by producing brass wire domestically, the vital English wool manufacture would be freed of its dependence on foreign suppliers of wool cards.[40]

Humfrey's plan to incorporate the manufacture of finished copper goods within the Company of Mines Royal soon encountered what he interpreted as active and underhanded resistance on the part of the Germans. The German mine managers' frequent complaints to London regarding the scarcity of wood to fuel their smelting furnaces had sparked his suspicions. Copper smelting was indeed a fuel-intensive process, and this was especially true for the "firm" English ore, which sometimes required up to twenty firings in order to extract and purify the metal.[41] English timber was becoming increasingly expensive during Elizabeth's reign as the demand for it rose, and most of the local landowners in Cumberland, who had opposed the mining works almost from their inception, had refused to sell their woodlands to the company except at "marvelous unreasonable" rates.[42]

Yet Humfrey believed that far more than convenient fuel supplies was at stake; the Germans, he wrote to Cecil in May 1565, were planning to use the alleged lack of smelting fuel in England as an excuse to transport the unprocessed copper ore overseas, where they claimed wood was more plentiful. After exporting the raw ore, the Germans planned to settle honestly with the queen and the English partners, compensating them for the value of the copper metal contained within it; yet they also knew as well as Humfrey did that the *unworked* metal, as the cost basis for their settlement, was worth comparatively little. Once the ore had been purchased outright from the English partners, the Germans would then be free to transform it into marketable copper and brass products themselves, using their monopoly of craft skills in battery and wire pulling. They could then sell the goods at a huge profit, which the English partners would never see; the naïve English would not even "think themselves wronged," Humfrey wrote, "although the best part and greatest profit to Her Highness, the realm, and to every particular partner of the English be carried away in a mist."[43]

When Cecil asked Humfrey how he had learned of this alleged scheme of the Germans, Humfrey confessed that he had discovered it while looking through their correspondence. He even told Cecil that he would be willing to act again in the future as a sort of corporate spy—a private, covert complement to Thomas Thurland's official public role as the English mine manager at Keswick—in order to keep the Germans from taking advantage of the English shareholders, as he was convinced they would try to do. "I shall see and understand what they do, and as occasion shall procure I shall report," Humfrey wrote to Cecil, for "if by some means craft be not repressed, [there] will be little plain dealing among some of them in mineral affairs." His patriotic desire, as he later explained, was to "make a sure defense for the queen's majesty and this realm," protecting the interests of the English partners from "all those caterpillars of the world which are collected into fellowships to recover treasures from all princes and commonwealths."[44]

Humfrey did not limit himself to spying on the German mine managers, however; such a passive approach might prevent the English partners from being swindled but would bring them no closer to obtaining the valuable expertise they needed. If Hechstetter and his German colleagues were unwilling to share with the English their manufacturing skills and the profits resulting therefrom, then Humfrey planned to find other Germans who would, and organize a rival venture. He somehow managed to establish a working relationship with a German artisan named Christopher Schutz, who agreed to introduce into England the arts of copper battery and wire pulling; Schutz even bonded himself to Humfrey for the sum of ten thousand pounds, should he renege on the agreement. According to Humfrey, Schutz was one of the most gifted and respected copper craftsmen in all of Germany; at one point he was supposed to have had charge of over one hundred laborers and answered directly to the Elector of Saxony, all at the tender age of sixteen. He claimed to know all the skills and techniques that Hechstetter and his people knew, and a few new ones besides. Moreover, he was willing to emigrate to England, become a subject of the queen, bring over a number of his assistants, and (perhaps most importantly) teach his arts to English apprentices, a number of whom Humfrey had already recruited for the purpose.[45]

Schutz's valuable expertise was not easy to come by, as German copper batterers in particular tended to be united in maintaining the secrecy of their art. Nuremburg was a great center for copper battery, and German merchants sometimes took their unworked copper metal there from a distance of more than seventy miles to have it pounded into pots and kettles. Even the towns in the immediate vicinity had not been able to acquire the secrets of this lucrative manufacture. Schutz claimed that the whole town where he had learned his art was "watched day and night like a town of war, to keep the said art secret to the first finders." Humfrey even wrote to Cecil that he and Schutz were afraid for their safety, "already fearing casualty of death" for their efforts.[46]

Others within the Company of Mines Royal were also aware of the need to turn their unworked copper into finished goods if the company was to prosper. Thomas Thurland had hoped that the Germans already employed in England might be induced to share their manufacturing expertise with the English, but Johannes Loner refused to be bound to do so. The German partners proved equally unwilling to help the English mine their own zinc, a necessity if the company ever hoped to produce brass domestically. Humfrey even complained to Cecil in November 1565 that Hechstetter had told him he knew there was zinc ore in England but would not tell him where, "for he is very secret, and so is Hans Loner, and by all the tokens that ever I could gather they mean . . . to keep from this realm that knowledge of battery, as much as in them lieth."[47] Humfrey, however, through his German associate Christopher Schutz, offered to introduce the same rare, lucrative, and highly prized arts into England, and even to teach them to native Englishmen. All he asked of the queen in return was a corporate monopoly on their use.

Humfrey soon left the Company of Mines Royal, on bad terms with Thurland, Hechstetter, and the other managers at Keswick, and had his revenge for the poor treatment he felt he had received at their hands. In 1565 he obtained royal permission to found his own joint-stock company, the Company of Mineral and Battery Works. The royal letters patent creating the company gave Humfrey and his partners the exclusive right to mine zinc ore in England; a monopoly over all mining rights besides the limited ones granted to Thurland and Hechstetter in 1564; and the exclusive right to exploit the arts of copper battery, brass manufacture, and brass and iron wire pulling. In effect, though the Company of Mines Royal had a monopoly on the production of copper metal (in certain counties), Humfrey had obtained a monopoly on all the ways of making copper marketable. For the rest of the sixteenth century, the Company of Mineral and Battery Works concentrated almost entirely on zinc mining and the manufacture of iron and brass wire, selling its wares to a large and stable market, usually at a respectable profit.[48]

Humfrey's departure did not restore peace to the company, however, for he was not alone in his suspicions. By 1565 several of the company's English shareholders had started to wonder what had become of all the money they committed to the venture. Partnership in the Company of Mines Royal was very expensive—the price of a single share was roughly £1,200, and this guaranteed only admittance to the company. Once a partner had purchased his share, he could expect to be called upon to make further financial contributions toward the mounting costs of labor and supplies, before pure copper metal could at last be produced and sold. The extra assessments upon each share were frequent and burdensome and were an understandable cause for complaint on the part of the English; by 1569 each shareholder had been assessed a further £850 to support the company's expenses, during which time no profits were ever shown.[49]

The enormous expenses were unavoidable; starting up a new mine was always costly, and profits might still be months or years away even if the mine was a rich one. The construction of dormitories for the laborers and private residences for the mine managers in the Cumberland wilderness; the digging, reinforcing, and draining of tunnels and shafts; the erection of smelting furnaces, stamp mills, and other complex machinery; and finally the digging and smelting of the ore itself were all obligatory but labor-intensive activities, and the managers had to have a great deal of cash on hand to pay the workers. The timber for building the mining community's facilities and fuel to run the furnaces were also expensive. Every one of these outlays was necessary before the mine could produce metal, cover expenses, and yield a profit. As experienced miners, managers, and investors, the Germans involved with the company all understood this.

Their English cohorts, however, did not. Most of them were accustomed to investing in trading enterprises, which despite their inherent risk had a much faster rate of return; they had little experience with or patience for the sort of long-term, capital-intensive type of investment that necessarily characterized new mining ventures. From the English perspective, all the partners had to show for their minimum twelve-hundred-pound investment after more than a year were the unfinished foundations for a smelting house, some large piles of rocks and dirt, and some very deep holes in the English countryside. They understandably expected to see some evidence of a return on their prior contributions before they committed any more of their money to the venture. As a result, by 1566 a number of the English partners were in arrears over the supplemental payments assessed to each shareholder. Even such a prominent figure as the Earl of Leicester was negligent in meeting the successive claims made by the company upon his two shares, a great disappointment to the German partners, who had hoped that someone of Leicester's wealth and stature might lead by example.[50]

The defaults of the English shareholders aroused the resentment of the German partners in Augsburg, who were doing the best they could to carry the debts of the discontented Englishmen and keep the company solvent during a vulnerable stage in its development. By the middle of 1566, the Germans were carrying roughly two-thirds of the company's expenses, while they still nominally owned only one-third of the company's shares. In April 1566 Hechstetter and Loner wrote from Augsburg to Lionel Duckett, a prominent London merchant, an alderman of the city, and the governor of the Company of Mines Royal. As governor, it was Duckett's responsibility to see that all the English partners' shares were paid up in good order. Hechstetter and Loner urged Duckett to be more aggressive in obtaining payment, using Cecil's help if necessary, "for in the works of the mines there must be no want of money." In Germany, they wrote, where mining matters were better understood, such problems did not exist, which was a key factor in allowing their mines to be profitable: "For if here in Deutschland there were not better helpers than diverse of our company be, and also better keepers of orders, there would not be so much copper and silver found out as there is."[51]

The tension between the German and English partners only increased with time. In June 1566 Hechstetter and Loner wrote directly to Cecil, alarmed that so many of the English were still in arrears. The German laborers at the mines needed to have their wages paid, and the mining works would founder without fresh supplies of fuel, timber, food, and other necessary supplies; all of this required ready cash, and the German partners were weary of having to fill the gaps created by the Englishmen's delinquency. They had written repeatedly to Duckett over the previous weeks and believed that he was sincerely doing his best, but they wished that Cecil would use his political weight to help him. Perplexed and frustrated by the recalcitrance of the English shareholders, and resentful of the implied suspicion of their competence and conduct, they lamented, "[W]e cannot judge the cause of their evil payments, but only that we think they mistrust us."[52]

Hoping to address that perceived mistrust head-on, Loner and Hechstetter sought to prove that they had not spent the company's money "for our [own] particular use and gains." They enclosed in their letter to Cecil a brief account of the company's considerable expenses to date: even with the extra payments assessed to each share up to that point, the company was still over twelve hundred pounds in debt. Meanwhile, their expenses continued to mount; if all the shareholders were not assessed another one hundred pounds very soon, the laborers would have to be discharged and sent home. This expensive step, once taken, could well have been a death blow to the company, for the German workers already felt ill-used in England. They knew they could earn the same or better wages at home in Germany, and once they left, they would be very unlikely ever to return. Loner and Hechstetter also pointed out that since the German partners were carrying even more than their fair share of the company's expenses, it made no sense to think that they would somehow be making money while the English lost it; all the company's shareholders were investing in the same project together, after all. Thus far, the two Germans insisted, everything at the mines had gone according to plan. The ore samples they brought to Augsburg had been assayed and had proven rich in copper. With the works already "brought to a good beginning with our labor, diligence, and charges," all that was needed for success was patience, perseverance, and courage on the part of the investors.[53]

Yet relations between the English and the Germans continued to deteriorate. Hechstetter returned to Keswick from Augsburg in the fall of 1566, appalled at the lack of order and discipline he found there. The mines themselves were not a problem; the miners had discovered several new veins of ore, many of which proved to be exceptionally rich in copper. Moreover, the builders had almost completed construction of the smelting house and were about to begin work on the furnaces for it, so that Hechstetter could report to Loner (who had remained behind in London), "forasmuch as I have yet seen, I do like all things very well."[54] But if Hechstetter was pleased with the progress his men were making in the mining works, he was dismayed at what he perceived to be the administrative incompetence and apathy of the English mine managers.

The English inhabitants of Keswick had violently attacked the German miners on several occasions during Hechstetter's absence, allegedly at the instigation of an individual referred to only as "that naughty man, Fissher," and one German named Leonard Stoultz had apparently been beaten to death. Hechstetter communicated his understanding of the incident to Loner: "[Stoultz] defended himself a long space [of time] against twenty of them, until the son of John á Wood struck him upon the arm with a staff, [so] that he could not any longer lift up the same [arm] for his defense, and then they fell all upon him and piteously murdered him, the chief occasion whereof was the said Fissher." Although many of the local English authorities were appalled and wanted the culprits brought to justice, the latter were protected by Lady Radcliffe, a member of the Cumberland gentry who held the lordship of the village of Keswick. Hechstetter called her "a great bearer of these seditious and naughty persons" and said that through her influence they were "set at liberty, and as ready to begin the like as ever they were."[55]

The situation at Keswick in late 1566 was tense, requiring firm discipline and prudent leadership; but Hechstetter had very little confidence in the English managers who had been left in charge there. Although he conceded that some did do their best "to maintain quietness and peace," they had very little help from most of their countrymen. Even Thomas Thurland, "being so much in fear to displease in anything the people of the country," had proved to be a disappointment as a manager, in Hechstetter's opinion: "[H]e entreateth them [the local villagers] gently and friendly, thinking thereby to do much, [but] they do abuse him with fair words and encourage themselves to more mischief. Our people have had a good opinion of Mr. Thurland, but seeing the things fall out as they do, a great number have altered their minds therein." If the "mischief" of the locals was allowed to continue unabated, Hechstetter wrote, it would be increasingly difficult to keep the German miners in England, since "they will not hazard their lives in such sort, considering that they have no more wages here (with this danger) than they might have at home with quietness."[56]

Their failure to maintain order and discipline, however, was only one of Hechstetter's complaints about the English managers; in his opinion, they had been no better at actually running the mines. Although the works had progressed notably during his absence, so that Hechstetter did "like all things very well," he apparently attributed this success to the merits of his own men, and not the administrative talents of the Englishmen nominally in charge of them. Thurland in particular seemed to Hechstetter to be simply too ignorant of mining works to do his job adequately; his good intentions were not in doubt, but his abilities simply could not match them. Hechstetter even suggested to Loner that "some more fit person might be by the company appointed in Mr. Thurland's room . . . for surely (although he would gladly do well) through his simplicity and lack of such understanding as appertaineth, he doth sometime more hurt than good."[57]

The partnership between the English and the German shareholders was crumbling. The English neither trusted the Germans nor were able to monitor their activities satisfactorily so as to restore their peace of mind. The most mineralogically competent Englishman who could be found, William Humfrey, had botched his assay of the Cumberland ore and conceded his inferior expertise relative to the German assayers. Afterward, he had become so openly suspicious of and antagonistic to the Germans that he left the company altogether to start a rival venture. Thomas Thurland, the man trusted by the English partners to protect their interests at the mines, did not appear to have the mining or managerial skills needed to handle the job and had succeeded only in irritating the Germans. The entire situation was exacerbated by the "economic nationalism" on the rise in western Europe during the sixteenth century, particularly in England.[58] Despite their partnership in the company, the Germans still represented foreign competition for English economic interests, and as outsiders they seemed most likely to stick together if questioned or confronted—a suspicion that Humfrey openly fostered and Thurland could not allay. The English partners, mistrustful of the foreigners who possessed an exclusive expertise they could not control, began to fear that their large investment was wasted, and they were very reluctant to risk further money in the venture. The German partners, who could not understand the profound mistrust of the English toward themselves, were offended by their refusal to honor their agreement and by the Keswick inhabitants' open hostility toward the German workers.

It was in the midst of such strained relations that Thomas Thurland, who had once praised his German colleagues as "wise men, true dealing men, and artificial men," at last came to fear that he too was being duped by crafty foreigners who knew far more about mining than he did. The atmosphere of mutual suspicion and the growing breakdown in communication between the two sides prompted him to write his desperate letter to Cecil in May 1566, complaining that the German miners seemed to be withholding valuable information from him in the absence of their own officers. Thurland's grave responsibilities at Keswick and his own inexperience with mining works made him aware of his vulnerability with respect to those who possessed an expertise he did not share.

With nowhere else to turn for help, Thurland appealed to Cecil to write in his behalf to Thomas Gresham, a respected and influential London merchant who made frequent trips to Antwerp during the 1560s and often undertook the Crown's business while he was there. Thurland asked Cecil to speak to Gresham, as secretly as possible, and charge him "to get me hither out of Flanders, as secret and as soon as he could, some expert and skillful man in minerals or some cunning goldsmith, by whose help and knowledge I might understand the certainty of those ores. And if it might be, I would wish him to be such a man as were both trusty and skillful, and had the Latin and the High [German] tongue, and yet were no [German] born." Thurland wrote further that he had already concocted a cover story to tell the German miners, so that they would not question the arrival of the newcomer or suspect him as a spy for the English. So strongly did he feel about the necessity of resorting to such a measure that he even offered to pay the man's travel costs out of his own pocket. He begged Cecil to keep the matter solely between the two of them and Gresham, lest some rumor reach the Germans and warn them of the spy's imminent arrival, causing them to cover their tracks or else provoking them into taking serious offense and departing England for good.[59]

Thurland never managed to bring any spies across the English Channel to help him monitor the German miners, but his very attempt to do so indicates his awareness of the uncomfortable situation he was in. Thurland's job, the only reason for his extended presence in such a remote corner of the realm, was to represent the economic and administrative interests of the English shareholders in London. He was expected to act as their chief agent, to ensure that the German miners were competent, their managers well-meaning and honest. The Englishmen's concerns were rooted in their self-conscious ignorance of, and hence their inability to control, the whole of the company's mining, assaying, and smelting operations. Aware that they could not catch the Germans in any hypothetical underhanded dealing, the English were susceptible to suspicions that they were being cheated—through an investment controlled by foreigners, with no end of expenses, and with no immediate returns in sight.

Thurland, then, was appointed by the English partners to serve as a trusted mediator, their very eyes and ears, to relay to them exactly what was going on at distant Keswick and so place the Germans' expertise under their control. His mediation was usually unsuccessful, because Thurland was scarcely more qualified to observe the Germans in the mines than his employers were; he had never even seen a profitable copper mine, let alone worked in one. In short, Thurland was out of his depth—he could not manage what he did not understand, and his impotence was most evident in his pathetic letter to Cecil, pleading for nothing less than a spy who possessed German mining knowledge but was "no German born." Communication had broken down so badly that the English were reduced to considering covert tactics, keeping secrets from the partners they feared (but could not prove) were keeping secrets from them.

## Cornelius de Vos and the Search for Scottish Gold

Thomas Thurland's mission at Keswick was not a complete failure, however. His mediation, though generally ineffective, was pivotal in allowing the English shareholders and German mine managers to work together when the company's well-being was threatened from without. In October 1566, shortly after Leonard Stoultz's death and just before Hechstetter's return from Augsburg, Cornelius de Vos arrived at Keswick in company with a Scot and an English merchant (neither of whom is named in the records). De Vos, a Dutch émigré, was included among the English shareholders in the Company of Mines Royal. He appears to have been something of a mining entrepreneur, with financial interests in several mining-related activities, including the production of alum and the draining of flooded mine shafts. In 1566 he had been prospecting for gold in Scotland, at a place called Crawford Moor, roughly eighty miles north of Keswick on the banks of the Firth of Clyde in Lanarkshire. De Vos was well informed in his choice of prospecting sites; gold had first been discovered in modest quantities at Crawford Moor in 1511, and the Scottish Crown had awarded various monopoly grants during the sixteenth century to those willing to try recovering the ore at a profit. Indeed, Daniel Hechstetter's father, Joachim, had been the recipient of one such grant in 1526, but within a few years he had abandoned the work and returned to the Tirol.⁶⁰

Despite his involvement in diverse mining affairs, De Vos shared Thomas Thurland's problem, in that he lacked the mineralogical expertise to determine for himself whether he was finding anything of value in Scotland, and he needed some outside help. As a shareholder in the company, De Vos was obviously aware of the mining operations at Keswick, and he also knew that if anyone in Britain could tell him whether or not the soil in Scotland contained gold, it was certainly the German assayers at work there. He told the Germans upon his arrival at Keswick that Scottish prospectors were washing the soil at Crawford Moor for gold and were throwing away large quantities of sand in the process. Suspecting that the Scots did not know what they were doing, de Vos collected a sample of the cast-off sand, brought it with him to Keswick, and asked the Germans assay it. The Germans agreed to perform the assay, and after doing so they informed him that the sand appeared to be very rich in gold. Once he had the information he came for, de Vos told the Germans that he could acquire ten thousand marks' worth of the sand by Christmas if they would be willing to extract the gold from it. He then returned with his companions to Scotland.⁶¹

For a shareholder in the Company of Mines Royal to use the company's resources and personnel in pursuit of a separate mining venture outside their original patent was certainly unusual, though given the political clout of the other shareholders, the patent could probably have been modified if necessary. What made De Vos's proposal a dilemma for the company's officers at Keswick was that he was apparently acting not in the company's, but in his own private, interests; De Vos hoped to win the Scottish mining rights for himself, and he did not want the rest of the shareholders to be involved. This meant that he could not reveal to any of the English, especially Thurland, exactly why he wanted to consult with the German assayers or what they found in his sample. While at Keswick, therefore, he operated as much as possible in secrecy. Although he apparently told the German assayers everything, in order that they might make a proper assay, he shared only bits and pieces of his story with Thurland and did not reveal the positive results of the assay to him.⁶²

Thurland may have known relatively little about managing a copper mining operation, but he recognized suspicious behavior when he saw it. He reported De Vos's visit at Keswick to William Cecil in a letter written 7 October 1566 and enclosed a separate, more formal report to be delivered to the queen herself, at Cecil's discretion. In his letters, Thurland reported that De Vos had brought with him to Keswick "certain sand in a napkin" and "conferred here secretly with our workmen to have a proof of it that I knew not of." He refused to be outmaneuvered, however; turning De Vos's tactics against him, Thurland managed to steal a sample of the sand without De Vos's knowledge, in order to assess its contents for himself.[63] But as he had explained in his earlier letter to Cecil requesting that a spy be sent for to help him, Thurland had no way to test the sand for gold on his own; with only his own minimal mineralogical knowledge to rely upon, the sand was quite as much a mystery to him as it was to De Vos. To obtain the answers he needed, he too would have to consult with the German assayers.

Surprisingly, perhaps, given the strained relations and atmosphere of mistrust between the German and English shareholders at the time, the German assayers came to Thurland's rescue. Rather than keeping the true purpose of De Vos's visit a secret, as he had intended them to, the Germans with whom he had consulted went to Thurland and told him the whole story. Their discretion seems to have restored Thurland's faith in them; he referred to them as "my strangers" and reported to the queen, "I judge [they] will keep no secret from me." Thurland then directed the Germans to repeat their assay of the sand "before my face" and found indeed that "it holdeth so much in gold as is more mete for a prince [to own] than any subject." Moreover, Thurland surmised (probably as a result of De Vos's suspicious visit and secretive behavior) that only a very small number of people could possibly know about the Scottish gold, because the only men in Britain who could verify its existence were the German assayers at Keswick. The Scots themselves were apparently still discarding the sand in question, believing it to be worthless. Thurland hoped that by taking full advantage of the Germans' virtual monopoly of mining and mineralogical expertise, the Company of Mines Royal might successfully bargain for the prospecting rights at Crawford Moor before De Vos or anyone else could do so.[64]

Thurland's letter attracted Cecil's attention; he wrote back to Keswick asking for more information and advice as to how the company's shareholders should proceed. During the intervening period, however, Thurland had had a chance to confer with Hechstetter (who had just returned to Keswick from Augsburg), as well as the other "expert men" who had first tested De Vos's ore sample. On 5 December 1566, Thurland reported to Cecil that the Germans had grown suspicious of De Vos and his ore. While confirming that the sample was indeed rich in gold, the Germans now told Thurland that "they much fear the true dealing of Cornelius therein" and suspected that he had seeded the sand with gold filings before the assays. Thurland suggested that the company might settle the issue by circumventing De Vos altogether and sending one of the German assayers to see Crawford Moor for himself and bring back his own ore samples, "and all Scotland never the wiser." Removing De Vos as a middleman would allow the company to make the fullest use of their control over German mineralogical expertise, keeping themselves fully informed while denying valuable information to potential rivals.[65]

The Germans encouraged this approach by assuring Thurland that the company's bargaining position for the Scottish gold prospecting rights was unassailable. Not only did they control the requisite knowledge and skill of the German assayers; they also controlled access to the rare expertise needed to wash and refine the gold. Hechstetter even promised that De Vos and the Scots would never be able to find skilled laborers to work for them, because the only ones who knew how to perform the required operations were in the Tirol and were thus already employed by Haug, Langnauer, and Company in Augsburg. In answer to Cecil's request for further advice, therefore, Thurland suggested that Elizabeth should approach Scotland's Queen Mary on behalf of "certain expert Germans in mineral affairs, who hath taken her Grace's mines here in England." She could ask Mary to allow the same Germans to carry out a survey of the lands at Crawford Moor and, if the soil there should prove to be rich in gold, to lease the prospecting rights to them.[66]

The Company of Mines Royal ultimately failed to secure the mining rights at Crawford Moor; instead, the Scots awarded Cornelius de Vos a nineteen-year monopoly over the prospecting rights, in exchange for a percentage of whatever gold he recovered. De Vos sold his share in the Company of Mines Royal and established his own joint-stock company in Scotland to fund the project in 1568. As Hechstetter had predicted, however, De Vos still could not independently acquire the mineralogical expertise he needed to exploit the Scottish gold washes. In October 1568 one of De Vos's colleagues, a Dutchman called Rennier, appeared at Keswick and approached Hechstetter and Loner for an assay of more Scottish soil samples. He told the Germans once again that if they found the samples to be valuable, he could transport to England "so much thereof as [they] will require at a very small price." In the meantime Rennier asked that Hechstetter send "a couple of his skillfulest men" back with him to Scotland, "to help him to dig and to direct him in those works." He also asked Hechstetter not to reveal his proposal to anyone else in the company, but the English manager George Nedham reported to Lionel Duckett that Hechstetter "told me all things . . . and asked my counsel and advice what was best to be done therein." Nedham advised Hechstetter of "the nature and quality of Cornelius de Vos, the deceit and inconstancy of the Scottish people, and what peril it is to deal with them," and the Germans apparently declined De Vos's offer.[67] Work at Crawford Moor continued sporadically through the 1570s and gradually petered out altogether until a new monopoly was issued to different prospectors in 1594.[68]

Although the Scottish gold episode did not bring material profit to the company, it illustrated for the English shareholders both the effectiveness of the Germans' monopoly of mining expertise in Britain and the company's potential power in controlling access to it. The Germans working at Keswick were not only a singular source of mineralogical information; they were also the sole means through which one might profit from that information. Despite his own lack of expertise, in this case Thurland's mediation allowed the English shareholders in London to take advantage of the Germans' expertise, deploying it selectively for the good of the company. Moreover, this was not the only instance in which Anglo-German cooperation protected the company from an outside threat: in 1567 the company was embroiled in a lawsuit between the queen and the Earl of Northumberland, who challenged the Crown's right to grant copper mining rights on privately owned estates. Once again, Thurland's deft mediation allowed the queen's lawyers to rely upon the Germans' expert knowledge, of both smelting processes and Continental mining law, to build their case and undermine Northumberland's, allowing the Crown to win the suit.[69] Despite these cooperative successes, however, Thurland's growing rapport with the Germans at Keswick was not enough to overcome the suspicion and ill will festering between the English and German shareholders, and the company's fortunes continued to wane.

## Debts, Dissension, and Decline

The company's worsening financial situation did little to improve relations between the English and German partners during the 1570s. Although the Cumberland mines did yield a considerable quantity of copper, and even a small amount of silver, the shareholders could find no profitable way to dispose of it. Exporting unworked copper metal from England was illegal, according to a statute of Henry VIII, because of its potential use in making brass ordnance that might one day be turned against English ships. The queen had only a limited need for copper herself, and in any case she was entitled to a free one-fifteenth share of all that was produced, making her unlikely to purchase more. Just as William Humfrey had predicted in 1565, the company could profit only by selling finished copper and brass goods for the civil market; but ironically, Humfrey's break-away Company of Mineral and Battery Works had already secured a royal monopoly in manufacturing such products.

The shareholders of the Company of Mines Royal eventually convinced Elizabeth to purchase some of their copper at full market rates, to allow some more to be exported to Russia and France, and to loan them twenty-five hundred pounds at 8 percent interest (a good rate for the time) to help them along.[70] William Humfrey also agreed to grant the shareholders—many of whom also owned shares in his Company of Mineral and Battery Works—a license to produce hammered copper vessels for sale. Despite their earlier disputes and subsequent rivalry, Humfrey wrote to Cecil that he had no desire to see the Company of Mines Royal fail and was willing to consider a strategic partnership for the good of both companies. Besides, the Company of Mineral and Battery Works had not been exploiting its battery monopoly; it concentrated instead on wire pulling, from which it reaped steady profits.[71]

The Company of Mines Royal did manage to pay off some of its many debts, occasionally in trade for finished copper goods, but it could never find a steady source of income sufficient to cover its expenses and turn a profit. Even the limited income it did acquire was partly illusory, the result of a bookkeeping trick—the company's sales of unworked copper all came directly from its huge and growing stockpiles of already-processed metal. The gross income from these sales was thus offset by exactly the amount that the value of its stocks diminished, providing ready cash to pay debts but leaving a net profit of zero. In no year before 1580 did the company's copper output even cover the cost of its production, let alone yield a real profit.[72]

The English partners, already unhappy with the extra financial demands imposed upon them before 1570, had consistently refused to provide the company with additional capital, forcing their German associates to take responsibility for many of the English shares as well as their own.[73] They proved even less willing to contribute further funds once copper had been successfully produced but could not be sold. The German partners, who were more willing to support the early, capital-intensive operations they knew to be inherent in starting new mining works, stretched their credit and borrowed considerable sums of money just to sustain the enterprise.[74] By 1570 they had grown weary of carrying both their own and the Englishmen's debts. The directors of Haug, Langnauer, and Company wrote a series of increasingly frustrated and desperate letters to Johannes Loner in London, pleading that they were paying far more than their fair share and had even less to show for it than the English had.[75] Believing that his firm faced financial ruin, Hans Langnauer himself made it clear to Loner that if the English did not pay the costs of their own shares, the Germans would cease to prop the company up with "our own and other people's money." He made Loner and Hechstetter personally responsible for the situation and charged them with salvaging the partnership.[76]

Loner in turn appealed directly to William Cecil, practically begging for his help to keep the company from disintegrating under the weight of its mounting debts. After recounting the lengths to which the German partners had already gone to keep the company solvent, he suggested that it was high time the English took a turn, imploring Cecil to convince Elizabeth to buy more copper, and to pay for it in advance. Stressing "the good . . . affection, that I have always borne unto this country," he was nevertheless adamant about the demands of his German employers, "[w]hich if we cannot obtain, then the world may understand, that the fault is neither in us, nor lack in the mines that is taken in hand."[77] The English mine managers, themselves acutely aware of the company's precarious financial situation, confirmed Loner's claims. George Nedham wrote to Cecil in late 1570, and Lionel Duckett wrote in early 1571, urging Cecil to appeal to the queen for some financial assistance.[78] A third correspondent warned Cecil that if the Germans were not given satisfaction, they might very well take their rare expertise and go home. This would be a total disaster for the company, since "if [Loner] and his company break off, and leave the works, we have no Englishmen that have skill to take them in hand," and the probability of securing new German miners on favorable terms was very low.[79]

When Cecil charged, along with a number of his fellow shareholders, that "we of the English part, had no knowledge of the accounts of the mines, nor how our money is spent," the English managers assured him that this was due in no way to the duplicity of the Germans, but rather to persistent English negligence. The company's account books had all been translated from German into English, on Lionel Duckett's orders, starting in 1568; but they had apparently remained in Duckett's possession, unread.[80] In addition, the German managers had been asking for months for a competent Englishman to come up to Keswick and help them "husband" the works. The Germans believed that the local merchants were taking advantage of them, and they hoped that by sending an Englishman "of some countenance," bearing letters from the queen and Privy Council, to purchase their supplies, they might end the worst abuses.[81]

The English partners, however, remained unmoved; they continued to mistrust the Germans and began to question openly the Tirolean miners' expertise. Throughout the 1570s, they still refused to contribute further funds toward the cost of their shares, and by 1572 the Germans were so frustrated that they stopped their payments as well. Loner wrote to Cecil in March of that year that the German partners wanted all the English shareholders who were delinquent in their payments to be kicked out of the company without compensation. They saw this as the only prudent and fair course of action, "both by our own experience in this beginning, and also by the like order used in all other places where mines are." In uncharacteristically bold terms, Loner reminded Cecil that the Germans had never wanted to take on English partners in the first place, and yet they were now made to suffer those partners' endless suspicions and accusations. His frustration evident, Loner actually dared to scold one of the most powerful men in England for the bad-faith dealings of his fellow shareholders: "For first some of the company here [in England], not finding the mountains of gold they looked for, have me not only in suspicion, but also charge me with deceit, and that I and such of my friends as have here dealt in the cause have induced them into the works as it were of purpose to deceive them. And on the other side, my friends and partners in Germany take it very ill towards me that they, induced (indeed) by my persuasion to bestow and spend such a mass of money as they have done in this realm, in the furtherance of so good a cause, should be so unkindly dealt withal."[82]

The occasional, meager sales of unworked copper and finished copper goods prevented the company from having to shut down its mining operations altogether in the absence of financial support from either the English or the German partners, but this income was not enough to eliminate the company's debts. By 1579 the financial situation was so dire that Haug, Langnauer, and Company finally decided to pull out altogether, absorbing a considerable loss in the process and scattering their shares among various Augsburg merchants.[83] This imposed an even greater burden and risk on the English partners, who were still unwilling to commit further capital to the venture. In 1580 Daniel Hechstetter (now working directly for the English partners) proposed a new scheme to lease out the company's exclusive mining rights. This clever plan allowed the company's shareholders to maintain long-term control over their monopoly and pay off some of their debts, while eliminating the immediate financial demands upon those unwilling or unable to furnish more money to the venture. In the meantime, financiers with more capital to contribute might benefit heavily from taking on a greater risk for themselves.

Thomas Smith, the royal customer of London and one of the original English shareholders, led a small group of investors (including Hechstetter) who acquired the lease of the Cumberland mines. The group agreed to pay the queen her one-fifteenth share of all metal produced, as well as a one-ninth part to the company, and an annual monetary rent of £433 6s.8d besides. During the period of Smith's lease, the mines' operating costs decreased sharply, so that the mines began to operate at a small profit. By the end of 1586, the company had used his rent payments to pay off their loan from the queen and had even recovered a net gain of roughly

£38 per share. This trivial sum, however, represented less than 2 percent of the original investors' capital outlay over twenty years, at a time when the Crown regularly borrowed money at more than 10 percent interest per annum.[84] The first profits of the Company of Mines Royal, so eagerly anticipated, were at best a Pyrrhic victory for the shareholders.

The company's troubles during its first two decades had nothing to do with the quality of ore in the Cumberland copper mines, or with the ability of the company's German mining experts to deliver what they promised. The mines were full of metal, and the Germans proved that they could dig and smelt enormous amounts of copper ore even in unfamiliar surroundings and circumstances. The company did not prosper as hoped primarily because it failed to find a stable and profitable market for its product. No one in England had a pressing need for unworked copper, and the market price to be had for it could not cover the cost of producing it. Exports, though occasionally permitted, were too infrequent to provide much relief. And even after the company had acquired both the license and the expertise needed to produce finished copper goods, it still had trouble competing with the established markets and economies of scale that favored its Continental rivals.

Beyond its inability to dispose of the copper it produced, however, the company was also crippled by the fact that its English and German shareholders neither trusted nor cooperated with one another, a direct result of the Englishmen's lack of copper mining expertise. Having forced their way into a partnership with the Germans, the English soon realized that they were in over their heads, and it caused them great anxiety. Aware that their lack of knowledge put them at a disadvantage, they tried to take steps to overcome it but found that true mining expertise could not be so quickly or easily acquired. None of them had any experience of their own with copper mining, after all, and it simply was not the sort of thing one could get from reading a book. Although some of the English were certainly familiar with the scholarly, humanist-inspired treatises of Georgius Agricola, for example, this alone could never provide them with the kind of expertise they required. "I have heard that Mr. [John] Chaloner is well learned in Georgius Agricola as touching speculation, and therefore may talk or write artificially," William Humfrey cautioned Cecil, concerning a prospective English shareholder, "but lacking experienced knowledge by daily working, Georgius Agricola is a present medicine to make a heavy purse light."[85]

Without the sort of expertise they would need to run a mining operation themselves, the English partners were forced to rely entirely upon the German partners to provide it. Yet the introduction of German mining expertise created a fresh crisis of management for the English, who found themselves unable to monitor or control to their satisfaction the experts they had acquired, and thus they were unable to trust them. The Englishmen's ignorance had made the German experts necessary, but it also made them a threat, because there was no independent means to verify their competence and honesty. The Germans, for their part, perceived the Englishmen's mistrust as a mark of their ignorance at best and a slander upon their own character at worst. They could never understand why the English were so skittish about a mining operation that appeared to them to be so very promising.

The English partners had hoped to reassert their control of the company by demanding the installation of a parallel set of English mine managers, to monitor the German miners at first hand and report on their activities and progress. Yet the English managers found it nearly impossible to supervise and assess something that they did not understand themselves. William Humfrey was probably the most mineralogically expert Englishman of his day, and yet his botched assay of the Cumberland ore was an implicit concession that the English had no dependable means of verifying the Germans' claims. Even Thomas Thurland, who seemed to trust the Germans at Keswick and called them "true dealing men," came to doubt their integrity and to fear that they had found some treasure in the mines they would not share with him. Although Thurland did prove himself capable of managing the Germans' expertise for the English partners when the company was united against an outside threat, in all his time at Keswick he never succeeded in making their expertise his own. Without the authority conveyed through the possession and perception of expertise, all Thurland's efforts at mediation could not convince the English partners to trust the German miners as he did. The only men in England who could speak with the essential authority of expertise, in fact, were the very Germans who were the object of suspicion.

The dilemma of the Company of Mines Royal, then, was twofold: it entailed not only a lack of technical expertise but also an inability to manage the expertise of others. Without an adequate mediator, investors and administrators in London did not feel that they could control the experts they needed to carry out their projects, and their anxiety led to a festering mistrust. Other projects, however, met with greater success. When Elizabeth's Privy Council determined to rebuild Dover Harbor in the late 1570s, their administration of the project was centralized in London to an unprecedented degree. During the course of the project, the councillors learned to rely upon expert mediation to keep them abreast of the construction works and advise them on the best way to proceed. Our story shifts accordingly, from the Lake District of Cumberland to the white cliffs of Dover.

CHAPTER TWO

# Expert Mediation and the Rebuilding of Dover Harbor

> And as improper and impertinent it is for carpenters and shipwrights to make seawalls and ponds, as it is for makers of ponds and seawalls to build fair houses or make good ships.
> THOMAS SCOTT TO FRANCIS WALSINGHAM, 24 MARCH 1583

> Yet in discharge of mine own duty I have finally set down in writing not only the manifold absurdities, perils, and apparent errors of their plat and devices . . . but also remedies for every mischief. Leaving the rest to God and your honors, whose wisdom can and will, I know, discern between affectionate opinions and sound reasons, warranted with experience of all other artificial harbors . . . as well beyond the seas as at home.
> THOMAS DIGGES TO FRANCIS WALSINGHAM, 8 JUNE 1584

Dover Harbor represented a vital interest for England's economy and defense in the late sixteenth century. Its proximity to the Continent made it a convenient commercial port for trade with the markets of northern Europe, while the ready access it afforded to the English Channel made it a key haven for Royal Navy ships—few naval harbors allowed a more rapid response against piracy or invasion. With several harbors along the southeastern coast of England in a state of ruin by 1580, the Elizabethan Privy Council deemed it essential to keep Dover open and viable. The redesign and rebuilding of the harbor during the 1580s was hailed by contemporaries as one of the most successful domestic achievements of Elizabeth's long reign, a triumph of ingenuity over nature that was worthy of attention and praise. Reginald Scot, for example, in his 1586 chronicle of the project, called the new harbor "a perfect and an absolute work, to the perpetual maintenance of a haven in that place, being such a monument as is hardly to be found written in any record."[1]

Yet during the early phases of the project, in the late 1570s, such a positive outcome seemed far from likely. Once the privy councillors had resolved to rebuild the harbor, they faced several difficult decisions. For one thing, there was no obvious individual to take the reconstruction in hand; instead, a number of candidates quarreled with one another, vying for the project's chief offices. Given that each candidate had a different design and plan for construction, the choice of a director for the works could have an enormous impact on the project's expense, timetable, and chances for success. All the aspirants claimed to be skilled and knowledgeable in hydraulic construction, but the privy councillors had limited means for distinguishing the truly capable from the incompetent. Prior experience was not a very useful criterion, since the project had no (English) precedents in either type or scale. How could the privy councillors find the engineering expertise they needed in the absence of a professional engineering community? How were they to differentiate between valid and false claims to competence in the absence of such modern means as professional accreditation and licensing? How could the proposals of numerous self-proclaimed experts be evaluated before they were carried out? Who among the inexpert administrators of the project was best qualified to make these evaluations?

This chapter explores how the Elizabethan Privy Council attempted to locate, assess, employ, and manage those would-be experts who claimed to have the knowledge and skills necessary to rebuild Dover Harbor. Although they possessed little or none of the relevant technical knowledge themselves, much like the English shareholders in the Company of Mines Royal, the privy councillors still had to evaluate the technical skills of others upon whom they would have to rely. The Dover project serves as another example of the use of expert mediators as an administrative tool, and it illustrates the complex interactions between skilled local craftsmen, central administrators in London, and the experts who mediated between them. As with the copper mines at Keswick, the privy councillors turned to a small number of trusted advisers to manage the construction works for them and to serve as knowledgeable intermediaries between themselves and a large, complicated project they could not fully understand. With respect to Dover Harbor, this management strategy was an Elizabethan innovation; earlier renovations to the harbor, during the reigns of Henry VII and Henry VIII, had been devised and carried out almost entirely by local townsmen and mariners. Only during Elizabeth's reign did the Privy Council assume that the final authority for such a traditionally local project should reside with them in London.

The man the privy councillors chose as their principal expert mediator at Dover was the mathematician, astronomer, and member of Parliament Thomas Digges. Among several competing claimants, Digges alone possessed the aggregation of knowledge, experience, and political and social connections needed to play an active role at all levels of the project's administration. His importance as a mediator is best illustrated through the resolution of two bitter controversies that arose during the early phases of construction: the building technique to be used in making the seawalls, and the location and orientation of the new harbor mouth. Once the privy councillors had accepted him as their trusted expert, Digges acquired considerable power in bringing about consensus among the commissioners and officers appointed to oversee the construction works at Dover. His opinions carried great weight and often proved so decisive that his opponents found him virtually impossible to overrule; his rivals either yielded to his authority or left the project. In either case, consensus was maintained among the project's managers, and the work could go forward as a result.

## Dover Harbor, an Eternal Struggle: 1495–1580

The geographic circumstances of Dover are vital to understanding its troubled history as a port town. One of the oldest and best-known ports in all of England, Dover is nevertheless an unstable place to build a harbor. It is located in a small bay at the mouth of the River Dour, though the "bay" itself is really little more than a slight indentation in the English coastline, extending perhaps a quarter mile inland from the open sea.[2] The town of Dover lies in the valley carved by the Dour between two high chalk cliffs, Archcliff and Castle Cliff. The harbor dates at least to the Roman occupation, though at that time the river was probably tidal, allowing boats to take shelter between the surrounding cliffs.[3] Even without regard to sailing up the river, though, Dover's cliffs offer a fine haven from any storm blowing out of the north or west (as well as an excellent platform for observation and defense), and its proximity to France no doubt made it seem a fine location for a harbor.

The harbor is constantly threatened, however, by the prevailing tidal currents of the area. During a flood tide, the waters of the Atlantic Ocean flow into the English Channel from the southwest, carrying large quantities of sand, silt, pebbles, and shingle. Upon hitting Ireland, a part of the prevailing current splits off and travels northward around Scotland, eventually generating a secondary current in the Channel from the northeast, opposing the direction of the main tidal flow. Where the two currents converge, very near Dover Bay, an unpredictable series of eddies is formed in which the speed of the waters is slowed and their cargo of sand and pebbles is deposited on the floor of the Channel, creating innumerable shoals and sandbars such as the Goodwin Sands, near Deal. The ebb tide, moreover, flows out of the Channel more slowly than the flood tide flows into it, because of the funnel shape created by the opposing coasts; the water moves more slowly in the widening direction and is thus far less effective at removing sand than the flood tide is at bringing it. The problem is exacerbated by the fact that the prevailing winds of the area are also from the southwest, augmenting the flood tide and hindering the ebb. Every harbor on the southeastern coast of England is consequently doomed, and only a constant and vigilant effort can prevent it from silting up and falling into ruin; this happened to a number of sixteenth-century English harbors, such as those at Camber, Winchelsea, and Rye.[4] By 1600 only Dover Harbor survived between Portsmouth and London, the sole southeastern English haven offering refuge from sudden Channel storms and the fastest port of departure for the Continent.

The first harbor at Dover of which there exists any more than legendary account was located in a slight coastal indentation between the town wall at the mouth of the Dour and an outcropping of Castle Cliff to the northeast. Although this would have provided some shelter from the tides, it also left ships exposed to the prevailing southwest winds, and thus in danger of running aground on the beach below the cliff. Nevertheless, this harbor appears to have served as Dover's primary anchorage until roughly 1495, when a local priest named John Clark developed plans for a new haven. Clark was the master of a hospital near the town called Maison Dieu, which offered shelter to traveling soldiers, sailors, and pilgrims who needed temporary lodging. Little else is known about his life, and there is no evidence that he had any experience with hydraulic construction before his effort to build the new harbor. In any case, Clark proposed that the haven be moved from the northeastern to the southwestern corner of the bay, to a pool created by a natural spring located at the foot of Archcliff. The pool was sheltered from the tides by a small, sandy outcropping, which he augmented by constructing a pier out into the Channel from the foot of the cliff. Clark produced and executed the plans for these modest works himself, with the help of local mariners and possibly some financial support from King Henry VII. The resulting haven came to be known as Paradise, and though apparently successful, it was never large enough to accommodate all the ships seeking harbor there.[5]

Although it provided an adequate short-term solution, Clark's construction set in motion what was to become a much more intractable problem at Dover; for as the flood tide of the Atlantic ran into his new pier, the resulting turbulence created an eddy just behind it, directly in front of the entrance to the new harbor. The tidal current was thus retarded and deposited its load of sand and shingle precisely in the most inconvenient place, the mouth of the haven itself. Realizing the problem, but not the enormous difficulty of overcoming it, Clark simply extended his pier further out into the sea, hoping that this would push the offending sand out into deeper waters where it would do no harm. His efforts were in vain, however;

the tidal current was slowed still further, the silting grew even worse, and by 1533 the new harbor was all but ruined, as is evident from an urgent petition of that year from the town of Dover, begging for relief.[6]

Once again the people of Dover took charge of the repair works themselves, seeking only financial assistance from London. Another Dover cleric, named John Thomson, having "consulted with the chief and best mariners of the town," journeyed to London and approached King Henry VIII and his chief minister, Thomas Cromwell, with his own plan for restoring the harbor. Thomson apparently acquitted himself well at the royal court; he had a personal audience with the king, "who heard his suit with great favor, and debated with him about the contents of his plat." Henry then ordered Thomson to go back to Dover and subsequently return to London as soon as possible with "some of the best mariners or seamen of the town" for further consultations. After Thomson and four Dover mariners conferred again with Henry in London, the king gave them an immediate grant of five hundred pounds from his own coffers to get the works started. He also appointed Thomson surveyor of the works, and after Clark's death, bestowed upon him the mastership of Maison Dieu Hospital.[7]

Thomson worked on the harbor repairs throughout the 1530s and 1540s. His original plan was to build upon Clark's works, extending the old pier even farther into the Channel in the hope of pushing the sand out far enough so that it would no longer choke the harbor mouth. He eventually managed to construct two jetties, one on each side of Paradise, which partially succeeded in keeping sand from accumulating in the immediate area of the harbor mouth; unfortunately, they were not long enough to push it out into deeper waters. Instead, the added obstruction of the new jetties caused the area just beyond the haven to silt up on a much larger scale than ever before, eventually sealing off the entire bay in front of Dover all the way to Castle Cliff. By the time of Elizabeth's accession in 1558, the harbor was once again all but ruined; a contemporary drawing of Dover shows that Paradise had become inaccessible, as demonstrated by the large number of ships moored outside of it while the old harbor remained empty. Despite the many years and the thousands of pounds spent in building them, Thomson's works were very nearly the end of Dover Harbor.[8]

No further action was taken until March 1576, when after repeated complaints from the town that its harbor was once again utterly decayed, the Privy Council (acting on the advice of a special parliamentary committee) commissioned four "men of experience" to travel to Dover, survey the harbor, and devise a plan for its repair. This action represents an administrative departure from previous projects at Dover, inasmuch as the Privy Council immediately assumed responsibility for the repairs and brought in outside consultants, rather than relying upon the townspeople to come up with their own plan. William Borough, one of England's most renowned navigators and skilled mathematicians, served as the head of the surveying party; though he had sailed the Arctic waters around Norway and Russia for over two decades by 1576, he had little or no experience with hydraulic construction projects. The Privy Council instructed Borough to make a general survey of the area around Dover, consider appropriate sites for a new harbor, and provide plans and estimates to the councillors for their consideration. Borough submitted his proposal two months later, but his estimate of thirty thousand pounds was high enough to table discussion of the plan for another three years.[9]

In February 1579, the harbor suffered further damage from winter storms, and the impatient Dover townsmen appealed once again to the Privy Council (through William Brooke, Lord Cobham, the warden of the Cinque Ports), this time for permission to take action themselves. The council responded by asking Cobham to assemble "some special and choice men . . . to take view of the said harbor, and to consider substantially by what means and with what charges" it might be repaired.[10] The Privy Council then formalized the assembly and appointed its members to a long-term commission for the rebuilding of Dover Harbor, charging them with the task of deciding how best to proceed. The Privy Council intended for the Dover commissioners to act as its local administrative agents, managing the project in its behalf—much as the English mine managers had served the Company of Mines Royal. The councillors trusted the commissioners to oversee the day-to-day details of the project and expected them to report all of their deliberations, actions, progress, and expenses to London on a regular basis.[11]

The formation of the Dover commission was a joint process, with both local and royal officials playing important roles; Cobham was responsible for nominating the commissioners, whereas the Privy Council endorsed all of his choices and suggested some of its own candidates. The group was fundamentally local in composition, with ties to Dover or Kent being a prerequisite for formal appointment, though the Privy Council also included nonlocal perspectives by occasionally sending its own consultants to Dover and ordering the commission to consider those individuals' opinions. The commissioners were an impressive group of Kentish gentry with diverse strengths, "some of them marvelous expert in affairs and matters of the sea, some in fortifications, some having traveled beyond the seas for experience and conference that way, and to see the order of foreign seaworks and havens, and none [was] without singular virtues," according to Reginald Scot.[12] Although none had actually overseen any hydraulic construction works before, their combined skills and experience were probably as close to the mark as could be expected in England at that time. The commissioners included Lord Cobham, Richard Barrey (the lieutenant of Dover Castle), the sitting mayor of Dover, and several other prominent local landowners. William Wynter, surveyor and master of ordnance of the Royal Navy, also acted as a frequent adviser to the works, as did William Borough, though neither was officially included in the commission, most likely because they were not from Kent.[13]

Of all the members of the Dover commission, the most important choice was Thomas Digges. His father, Leonard Digges, had been a prominent Kentish landowner and was the author of several mathematical treatises on surveying, mensuration, and mathematical instruments. Leonard had been condemned for his participation in Thomas Wyatt's Protestant-inspired rebellion against Queen Mary, though his sentence was commuted and his lands restored upon payment of fines; Thomas and his brother were restored in blood after Elizabeth's succession and so were able to inherit the family estates. Thomas was a pupil and friend of the mathematician John Dee and was himself considered to be one of the most skilled English mathematicians of the sixteenth century. In addition to editing and publishing many of his father's mathematical treatises, he wrote several of his own, addressing practical subjects such as surveying, navigation, and fortification as well as more purely mathematical topics in geometry and astronomy. He is perhaps best remembered for his *Alae seu Scalae Mathematicae* on the 1572 supernova and *A Perfit Description of the Caellestiall Orbes*, a translation of the cosmographic portions of Copernicus's *De Revolutionibus* that included a graphical depiction of an infinite heliocentric universe. Beyond his mathematical activities, Digges was an active member of Parliament after 1572 and had participated in preparations for the Earl of Leicester's proposed military expedition to the Low Countries in 1578, when he (Digges) had personally toured and reported on the condition of Dutch fortifications.[14]

By 1579, then, Thomas Digges was at the height of both his political and intellectual careers and seemed an ideal choice to take a leading role in the Dover works. As the owner of large estates in Kent, he was well connected among the gentry there, and his service in Parliament had won him the respect and patronage of members of the Privy Council. Although he had no actual experience in harbor or seawall construction, his mathematical education had been practical enough to make him a skilled surveyor and cartographer, and he had almost certainly learned the basic principles of Dutch harbor design during his service in the Low Countries the previous summer. Moreover, he was already quite familiar with the terms of the Dover project, having served on the parliamentary committee that originally recommended William Borough's 1576 survey. Over the next five years, Digges's unique combination of skill, experience, and political and social connections allowed him to play a central role in the Dover Harbor works as an expert mediator.

The Dover commission's first challenge was to settle on a plan of action for the project and submit it to the Privy Council for approval. Because Borough's original estimate of thirty thousand pounds had been deemed unaffordable, the commissioners sought other proposals. Aware of the sophisticated hydraulic works being built throughout the Low Countries, they "sent over unto Flanders for sundry men of experience of that country" to come to Dover and survey their harbor.[15] The Flemings, who traveled to Dover from Dunkirk, devised their own plan for repairs and produced a much more agreeable estimate of sixteen thousand pounds; in the meantime, Borough also revised his original plan and lowered his estimate to a more manageable twenty-one thousand pounds.[16]

The designs used in each plan turned out to be remarkably similar; both Borough and the Flemings realized that the accumulation of the sandbar across Dover Bay, caused by Thomson's works of the 1530s, provided them with a novel opportunity. So much sand had built up in the bay that by 1579 it enclosed a standing freshwater lagoon in front of the town, fed by the River Dour. The old haven, such as it was, could only be reached at the point where the river water fought its way through the sandbar to the sea, a point that shifted constantly and was never very deep. Their proposal was to utilize the lagoon, transforming it into a permanent backwater "pent," or reservoir, by building seawalls along the sand and shingle already present. The seawalls were to consist of two parallel rows of wooden piles pounded into the sea floor, with the gap between the rows to be filled with tightly packed earth—a traditional Flemish technique. The pent would serve as a holding tank for a large quantity of river water, to be controlled by a sluice gate set in the new seawall. When the sluice was opened at low tide, the collected river water would flow out through a newly constructed harbor mouth with enough force to scour out the sand and shingle that had settled there. In theory, the new harbor complex would then be self-cleaning and could be maintained forever. Despite the Flemings' lower estimate, the commission decided in favor of Borough's revised proposal.[17] Apart from some slight differences of opinion regarding building materials, it seemed that the project's sole remaining hurdle was to acquire sufficient royal funding.

## Conflict, Confusion, and Stagnation: 1580–1582

Despite the consensus rapidly emerging among the Dover commissioners, however, in August 1580 the Privy Council overruled the commission's decision.

In London, a man named John Trew had submitted his own rival plan for repairs directly to the councillors, circumventing the local authorities charged with evaluating such proposals. Not much is known of Trew outside of his involvement with Dover Harbor; he had apparently developed a reputation for skill in hydraulic construction by building a canal linking the city of Exeter to the sea in the 1560s, although the patrons of that project were never fully satisfied with his work there.[18] In any case, according to William Borough he was "a man well known unto some of their honors [the privy councillors] to be very skillful in water works," and Reginald Scot wrote that he "made great show to be an expert engineer."[19]

The unconventional nature of Trew's proposal for Dover might have warned the privy councillors to be suspicious of his qualifications. Unlike Borough and the Flemings, Trew saw no need to construct a costly pent. He planned instead simply to reinforce and extend Thomson's piers, the works that had caused so much damage in the first place, with a massive wall of hewn stone, "the which he framed after a strange and contrary kind of workmanship," according to Scot.[20] Perhaps as a compromise, he also offered to construct a small pent within the town of Dover itself, though what scouring power this could have had so far from the harbor's mouth (a distance of more than a half mile) is unclear. The great appeal of Trew's plan for the privy councillors, however, was its limited cost; because his proposal did not include construction of the large pent, Trew's estimated expenses were less than half of Borough's. Swayed by his low estimate and his prior reputation, in August 1580 the Privy Council unilaterally appointed Trew as the first master surveyor of the Dover works.[21]

The privy councillors acted with some discretion initially; they ordered the Dover commission to contract with Trew, at ten shillings per day, to build only the first two rods (thirty-three feet) of his stone seawall, as a test of his methods.[22] But haste soon overcame caution. Within a month Cobham wrote to Francis Walsingham, principal secretary of the Privy Council, that Trew was unhappy with his limited contract; he wanted to begin building the whole wall in earnest, and he asked that his work be evaluated only after it was completed.[23] Just ten days later, the Privy Council ordered the lord chancellor to grant Trew a commission to take on the entire project as master surveyor, giving him the right to hire and fire laborers, purchase and requisition supplies, and even imprison anyone who hindered his work—all this before he had constructed so much as a single foot of his proposed seawall.[24]

Although Trew's activities are seldom mentioned in the record of the Dover works after this point, the rest of 1580 and much of 1581 were probably spent hewing stone for his massive proposed wall. He apparently never managed to lay a single stone, however, and Reginald Scot accused him of deliberately stretching out the process in order to make the most of his daily wage. By the middle of 1581, Trew had come under fire from Borough, Digges, and the rest of the Dover commission. When the Privy Council pressed him for a progress report and an account of his expenses, Trew declined to provide them, saying only that "he would make them a good haven," though he could not say when it would be completed.[25] By June 1581 even his supporters on the Privy Council had begun to doubt his competence; they dispatched a number of consultants to Dover to survey his progress, including William Wynter, Thomas Digges, William Borough, Richard Hakluyt, and Francis Drake, and also asked the Flemings to come back for another look. By July the councillors had given up on that summer's working season altogether and simply ordered supplies to be stockpiled for the next year. Trew was not officially dismissed until he had made a full account of his expenditures more than a year later, but at this point his contribution to the works was ended, "for that his work is insufficient."[26]

The unmitigated failure of Trew's plan, and indeed his very participation in the project, illustrate the managerial pitfalls arising from the lack of a satisfactory expert adviser and mediator. As was the case within the Company of Mines Royal, the central administrators in London could not bring themselves to trust the judgment of those whom they had selected to manage their affairs at the local level. Trew's unilateral appointment as master surveyor of the works invalidated the prior resolution of the Dover commission to proceed with William Borough's plan and made it abundantly clear that final authority for the harbor's repair resided in London, not in Dover. Moreover, the Privy Council's principal criterion for its decision was not the plan's feasibility but its economy; the councillors emphasized that Borough's proposal required "a far greater charge than can be yielded," while Trew promised them success "with a small charge."[27] Trew's proposal was never the most

---

*Figure 2.1.* A survey of Dover Harbor and Bay, 1581; north is located at the top. This survey was performed either by or for Thomas Digges, whose coat of arms is on the plan, shortly before he submitted his own proposal for fixing the harbor to the Privy Council. The map clearly depicts how the sandbar across the bay had cut off all access to the harbor from the sea, save only the shallow, shifting channels carved by the river. The two faint, perpendicular lines near the center of the plan are sketched almost exactly where the walls containing the pent were to be built. The numerous annotations throughout the harbor and surrounding waters indicate water depth and bottom condition. According to Stephen Johnston, this plan is an eighteenth-century copy of the 1581 original, on paper watermarked "J WHATMAN." "Making Mathematical Practice: Gentlemen, Practitioners, and Artisans in Elizabethan England" (D.Ph. diss., Cambridge University, 1994), 226. Photo used by permission of the British Library, Additional MSS, 11,815A.

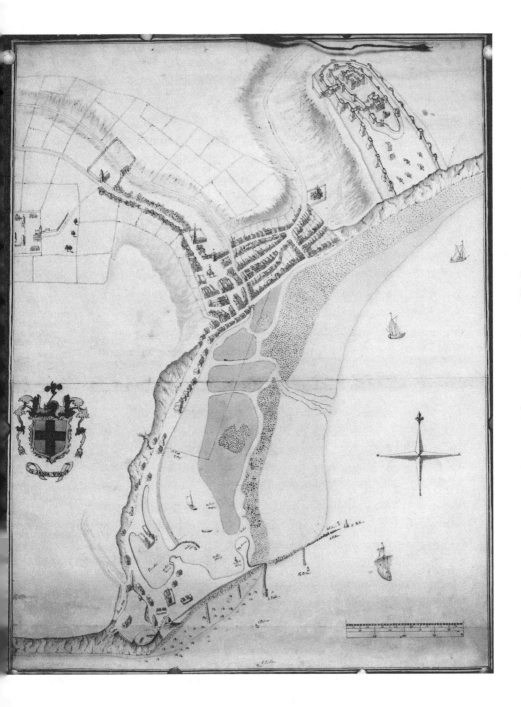

promising plan on the table; it represented merely the lowest bid—sufficient reason, nevertheless, for the Privy Council to disregard the consensus opinion of the commissioners and consultants they themselves had appointed and depend solely upon Trew's word. As a result, Trew was permitted to waste more than twelve hundred pounds and more than a year's working time.[28] Without an expert mediator whom they trusted, the Privy Council were vulnerable to such errors of judgment.

The competition to replace Trew did not take long to begin. Borough and the Flemings again offered their respective plans for consideration, and this time they were joined by Thomas Digges. Digges believed himself to be primarily responsible for demonstrating Trew's incompetence to the Privy Council, and he felt that this made him the logical choice to be Trew's successor. He had already made his own thorough survey of the harbor as it stood in 1581, and in early 1582 he submitted his plan for repairing it directly to the Privy Council. The heart of Digges's plan stuck closely to the Flemish/Borough model originally agreed upon by the Dover commission, though it involved some elaborate modifications, including rerouting the course of the River Dour. To support his case, Digges stressed his personal observation of harbor design and hydraulic construction in the Low Countries.[29]

At the same time, a man named Fernando Poyntz presented a plan of his own to the Privy Council, circumventing the Dover commission just as Trew had done before him. Little is known about Poyntz's life or career before his involvement with the Dover project. Like Digges, he claimed to have traveled widely in the Low Countries, and he had previously been commissioned to build some seawalls at Woolwich and Erith beaches along the Thames, just downriver from London, to keep the river's flood tide from inundating the area's surrounding farmlands.[30] Unfortunately, Poyntz's formal plan for repairing Dover Harbor has not survived; that he was able to impress the Privy Council with it is certain, however, perhaps in part because he offered to take financial responsibility for the works upon himself as a speculation. Although the Dover commissioners preferred Digges's proposal, the Privy Council overruled them once again in the spring of 1582 and ordered Poyntz to begin his works by draining the lagoon in front of the town and building some reinforcement jetties along the sandbar.[31]

The memory of Trew's expensive failure, however, may have made the privy councillors leery of entrusting the whole project to just one man. This time they hedged their bets by ordering the Dover commission to appoint Thomas Digges as the master surveyor of the project; Poyntz was to be the chief overseer at the harbor itself.[32] In addition, a third rival who had submitted some plans of his own directly to the Privy Council was hired as Poyntz's deputy: Thomas Bedwell, a Cambridge-educated maker of mathematical instruments.[33] The councillors may have hoped that by creating a situation of overlapping authority between Poyntz, Digges, and Bedwell they could rely upon each man to check the others. This would prevent any one of them from assuming full control of the project and thus guard against the sort of unchecked waste of time and money that had marked Trew's tenure as surveyor. They may also have believed that having several experts working on the project simultaneously would multiply their chances for a successful outcome.

Digges, who resided in London throughout the works, might seem an odd choice at first for the office of master surveyor, but then surveying per se was never his primary responsibility. Instead, his role was to be much like that intended for Thomas Thurland in the Company of Mines Royal: Digges's patrons on the Privy Council relied upon him to act as an expert mediator between themselves and the Dover commission. Digges was unique among those involved in the works in that he already had personal and professional relationships with both the privy councillors and the Dover commissioners, in addition to some knowledge of harbors and hydraulic construction in the Low Countries, as he repeatedly mentioned in his reports to the council.[34] As master surveyor, Digges advised the privy councillors in person on all aspects of the project, supplementing and explaining the commission's regular reports. His duties therefore compelled him to stay in London, though he did travel to Dover periodically to attend commission meetings and to observe and assess firsthand the progress Poyntz and Bedwell were making. Digges's ready access to the Privy Council eventually gave him a greater influence over their decisions than any of his fellow Dover commissioners enjoyed, although his required residence in London limited his practical contribution to the works while Poyntz was in charge at Dover.

The redundant management structure for the project caused trouble almost immediately. Poyntz quickly came under attack from all sides, but he lacked the power and authority he needed to defend himself. In May 1582 Digges presented to the Privy Council a comparative estimate of Poyntz's plan and the Flemish one previously considered by the Dover commission and concluded that Poyntz's works were less likely to succeed at more than double the expense. To make matters worse, Poyntz had failed to keep detailed records of his expenditures, claiming that he was not aware he was expected to do so.[35] Bedwell, too, was quick to blame Poyntz for all the project's failures, hoping perhaps to take over his office upon his dismissal.[36] Even Poyntz's friends on the Privy Council scolded him for failing to build some jetties to preserve a section of beach on which stood some of the queen's properties; he answered weakly that he lacked sufficient supplies and that the project ought to be left for later in any case.[37]

Poyntz clearly felt besieged by his many critics; he blamed lazy workers, bad advice from the townspeople, and poor weather for his difficulties. He argued that some of the Dover commissioners, especially Richard Barrey, the lieutenant of Dover Castle, were personally out to get him. He offered to resign if his efforts were deemed unsatisfactory, but the Privy Council refused to accept his resignation. Poyntz also had some defenders among the commissioners: Thomas Scott expressed his support, and the mayor of Dover championed Poyntz's cause to the Privy Council in strong terms, complaining of the "infamous libels of some lewd disposed persons set up against him."[38] The criticism of Poyntz was not altogether unwarranted, however; his efforts to drain the lagoon were only partially successful, lowering the water level by some two feet at best. His sluice works tended to break down in bad weather, and almost no one believed that they would survive the first winter storm.[39] Indeed, later plans of the harbor show that those of his works that were not incorporated into subsequent construction had decayed into ruins within fifteen years.[40] Poyntz had spent another thousand pounds on his feeble works, and another year had been squandered, with the harbor still no nearer to completion.

The Privy Council's continued reluctance to rely upon the judgment of its subordinates had given rise to a crisis of management between London and Dover. By the fall of 1582, Poyntz held the confidence of only a small minority of the Dover commission, but he could not be dismissed because of his supporters on the Privy Council. The simultaneous appointment of Poyntz, Bedwell, and Digges to positions of competing authority had created an impasse, since no one had the power and backing necessary for his own plan to overcome opposition from his rivals. The Dover commission did not have the authority to break the stalemate, since all of their decisions had to be approved by the privy councillors, who were inclined to disregard the commission's collective opinion. Instead of placing all their eggs in one basket, as they had with Trew, the Privy Council had placed a single egg in each of several baskets, resulting in such conflict that it seemed the entire project might be mired in stagnation indefinitely. Early in 1583, however, a solution was sought from a previously untapped source: the inhabitants of the nearby coastal town of Romney. The traditional Romney technique for building sturdy seawalls eventually attracted enough supporters, in Dover and then in London, to end the stalemate and get the harbor project moving once again.

## Digges, Scott, and the Romney Marsh Men: 1583

The main issue of debate during the winter of 1582–83 was the building material to be used in constructing the two seawalls enclosing the pent—the long wall (to be built along the sandbar, guarding the pent from the sea) and the cross wall (to be built from the long wall to the mainland, across the lagoon, and containing the master sluice). The integrity of these walls was vital to the success of the project, yet it was by no means obvious how best to build them. They would need to be sturdy enough to hold up against the regular tides and severe storms that can plague Dover; they would have to be built over an uneven, unsettled foundation, a mixture of bedrock, mud, and sand; they would need to contain a pool of river water at least twelve feet deep without allowing any water to leak through or underneath them; and they would have to be affordable, within the constraints of the large but limited budget of the project.

Fernando Poyntz had planned simply to pile up along the sandbar the very mud and sand to be dredged from the floor of the pent; he reasoned that if an existing sandbar was already holding in the lagoon, he had only to make the sandbar bigger for the lagoon to become permanent. Very few had any faith in this idea, but William Wynter supported it on the strength of its very low cost and Poyntz's reputation. The queen's master carpenters and shipwrights, Peter Pett and Matthew Baker, predictably advocated building the pent walls out of wooden planks; adapted from Flemish techniques of hydraulic construction, each wall would consist of two rows of planking with chalk and gravel used to fill the middle. This plan would have been costly, since timber was increasingly scarce and expensive in southern England during the late sixteenth century. Nevertheless, it had William Borough's support—it agreed with his original proposal, and perhaps, being a seaman by trade, he trusted master shipwrights to build sturdy, watertight walls.[41]

What would probably have developed into a protracted debate was just beginning when in March 1583 Dover commissioner Thomas Scott introduced a new idea for consideration. Having conferred with a kinsman of his from the neighboring town of Romney, he suggested that the seawalls townspeople had built there to prevent the ocean from flooding their arable land faced conditions very similar to those at Dover, including the uncertain foundation and the pounding of the sea.[42] The Romney seawalls were made of a mixture of earth and chalk (found in abundance at Dover, whose famous white cliffs are primarily composed of chalk), with a coating of mud beaten into the sides.[43] Unlike stone walls, they worked very well on uncertain foundations, since they were lighter in weight and actually became stronger and more watertight as they settled over time; they were also easy to repair if breached. They were very inexpensive, since they required almost no timber, and were made of materials available locally, in great quantities and at no cost. They had already proved to be sturdy and effective, and nearby there was an entire town full of laborers and overseers with ample experience in building them. Thomas Scott himself had helped to oversee the maintenance of the Romney walls for twenty-five years, playing a role not only in their construction but in keeping track of their costs as well.

This was not the first time that the Romney technique for hydraulic construction had been suggested, though it is difficult to imagine why it had thus far received so little attention. Thomas Digges was probably referring to the Romney method in his proposal of 1582, when he pointed out that "chalk with ooze [mud] doth singularly bind" when mixed and that "great cliffs and mountains of chalk" were readily available for free at Dover. This, he wrote, would save the project "infinite waste of timber, and endless charge of reparations" in comparison with Borough's plan.[44] An unsigned plan from roughly the same time, probably also composed by Digges, ruled out hewn stone as infeasible and timber as too expensive and asserted that seawalls might be constructed of "bavin artificially couched, interlaced with earth, ooze, beach, etc.," which could "firmly contain any mass or weight of water, and also be settled on any unsure soil."[45] Still another anonymous report on the harbor, written in March 1582, explicitly states that the walls of the pent should be constructed "by our marsh men . . . those workmen of Romney marsh."[46] Nevertheless, Scott's 1583 letter to Walsingham, and the minutes of a Dover commission meeting from the same time, represent the first extended consideration of the method. The commission minutes even included a diagram of the planned wall, with a cost estimate of eleven pounds per rod, as compared with an estimated thirty pounds per rod for walls made of timber planks.[47]

The Romney plan quickly aroused enthusiasm among the other Dover commissioners. Poyntz was an early, if tentative, supporter, perhaps in part because the Romney men reported that the lagoon would not need to be drained any further than it already had been (though Pett, Baker, and Bedwell all disagreed).[48] Edward Boys, another Dover commissioner, gave it his wholehearted approval, citing "the probability of reason, the demonstration of the like for walls in Romney Marsh, [and] the warranty of those honest, skillful marsh artisans." Thomas Digges wrote to Walsingham that the Romney men were "the only and fittest workmen" for the job, so long as they received proper guidance "for the form and some such other points as their skill and experience reacheth not unto." The positive assessment was not unanimous: mistrusting the Romney men's choice of building material, the carpenters Pett and Baker, together with Borough, still argued for a Flemish-style wooden wall. In the end, however, the Dover commissioners settled upon the men of Romney Marsh as the most expert men for the job. As Thomas Scott argued tellingly to Walsingham, with reference to the woodworkers Pett and Baker: "And as improper and impertinent it is for carpenters and shipwrights to make seawalls and ponds, as it is for makers of ponds and seawalls to build fair houses or make good ships."[49]

As in the past, the Dover commission's consensus was not in itself sufficient to warrant action; the commission still acted at the will of the Privy Council, and the councillors were still reluctant to trust the commissioners' judgment. This time, though, they were not so quick to ignore or overrule the commission as they had been in the past. Thomas Scott began a lengthy process of deliberation by sending the surveyor (his cousin, Reginald Scot) and the common clerk of the Romney works to London to confer with the Privy Council directly.[50] The councillors inquired about the Romney men's experience with building and maintaining such works, the soil conditions at Romney, the substance of the walls there, whether repairs could be made in the event of a leak, whether the walls could be built in standing water, and the likely cost per rod. The Romney men answered most of the questions as best they could, but they were vague about the estimate and, though they insisted that leaks could be easily repaired, like most skilled artisans they refused to disclose their craft secret for doing so.[51]

Thomas Digges, still serving in the office of master surveyor, championed the Romney method in London and once again was uniquely suited for the role. He had been one of the earliest proponents of the construction technique, having suggested something very similar a year earlier, based (he said) upon his observations of its successful use in the Low Countries. As a prominent Kentish landowner, he held a seat on the Dover commission and could claim firsthand knowledge of the conditions at both Dover Harbor and Romney Marsh. He could therefore vouch for the overall similarity between the two places and attest to the Romney method's efficacy, despite the refusal of its creators to reveal their craft secrets. During his years as a member of Parliament, meanwhile, he had earned the trust of the powerful privy councillors, who most needed to be convinced.[52] Relying on the same knowledge, experience, and connections that had won him the office of master surveyor in the first place, Digges could advocate the Romney method to the Privy Council more directly and effectively than any other member of the Dover commission.

In order to secure the trust of his patrons, an expert adviser and mediator such as Digges must not be seen to have his own agenda; he had to be (or at least, to seem) personally disinterested in the project at hand, making his patrons' interests his own. In advising the Privy Council regarding Dover Harbor, Digges always took great pains to stress his own objectivity and denied having any personal interest in the project so long as the harbor was successfully completed in the end. After all, he wrote, a viable harbor of any kind would increase the value of his estates in Kent, regardless of whose plan or what materials were used to build it.[53] Alone among the project's officers, Digges renounced the salary of ten shillings per day to which he was entitled as master surveyor, and he even paid his own travel expenses, telling the Privy Council that he desired "neither pay nor profit, nor other recompense, but only your good favor and honorable allowance of my long continued, painful, chargeable travails therein employed."[54] This contrasted sharply with Trew, who was accused by some of delaying the works on purpose in order to draw his salary as long as possible.[55] Digges also compared his own objective position with that of the carpenters Baker and Pett, who were pushing so hard for wooden walls, "for what private respects I know not," implying that the two stood to profit if their plan in particular was adopted.[56] Finally, Digges claimed to have refused a bribe of five hundred pounds "to hold my peace, and to give my looking on only," in the Romney debate.[57] His reported refusal emphasized to his Privy Council patrons his honest and disinterested commitment to the project, while the very claim that he (and no one else) was the target of a bribe implied further that his opinions held, and should hold, special importance.

Digges prevailed with most of the privy councillors in the Romney issue, winning over the most senior and powerful member (who was his personal patron), William Burghley, who stated that "if he [Burghley] erred therein, as not seeing but hearing the matter in question, he would err with discretion, as led by the reasons of the commissioners, who had seen and tried the experience of that kind of work."[58] Walsingham, however, proved more difficult to convince; he was moved in part by William Borough's dissension in favor of wooden walls, on the assumption that "walls of dirt" could never be made watertight.[59] Poyntz's plan simply to pile up mud and sand also retained a few adherents, because of its expedience and low cost. Poyntz still enjoyed a reputation as a "good engineer" with the councillors, according to Reginald Scot, "partly for his experience in foreign works, partly for his resoluteness: but especially for that he made a show of more cunning than he would utter." This was despite the fact that as late as 1586 his works at Woolwich and Erith beaches were "as yet unaccomplished, though no small charges have been therein employed."[60]

After extended discussions with Digges, Poyntz, Borough, and the representatives from Romney, Walsingham tentatively pronounced in favor of the Romney method, but he ordered another consultation with William Wynter, who asked all the same questions yet again. During this process of deliberation, William Borough eventually acceded to the Romney plan. Poyntz's idea was finally rejected, but Wynter and Walsingham maintained their confidence in him. They proposed to send Poyntz back to Dover, where he might be engaged in building some of the less difficult works on the Romney model, though in the end he elected to quit the works instead, complaining that his limited portion of the work was "needless" and ill-advised. Although his name appears in the record once or twice more, this marked the end of his active participation in the project.[61] The works went ahead without him; on 10 April 1583, with the Privy Council's approval, the Dover commission at last hired the laborers of Romney Marsh to build the seawalls "of greatest difficulty," namely the walls that would enclose the new pent.[62]

The Romney method translated very well indeed to Dover Bay, and the works of 1583 were a resounding success. As an early (and prudent) test of the method, the entrance to the spring-fed pool that had previously served as the old harbor, Paradise, was sealed off and regulated by a new sluice. Within days the sluice had broken, forcing the new wall to hold back tons of water, yet no leaking was observed. Once repaired, the new sluice helped in scouring the harbor mouth, kindling great hopes for the master sluice to be placed in the pent's cross wall.[63] The commissioners carefully chose the main period of construction to fit between sowing and the harvest, a time when idle farmhands were more of a financial liability than an asset; most local landowners were happy to have their tenants earning extra money elsewhere for the summer. More than a thousand laborers in all were employed at the works, some traveling more than ten miles to get there, and in the end some had to be turned away because there was not enough pasture land for their cart animals. So many laborers responded to the request for help, in fact, that Thomas Scott (replacing Poyntz as the head overseer) decided to begin work on the long and cross walls simultaneously; he assigned Richard Barrey to oversee the long wall and took on the more difficult cross wall himself.[64]

The laborers worked very long hours throughout the summer, hauling enormous quantities of mud and chalk out into the bay using wooden carts. Sometimes the work was dangerous; Reginald Scot wrote that several men toiled all night in chest-deep water, for example, when the time came to install the master sluice in the cross wall. They enjoyed excellent weather, however, and the work proceeded rapidly with no casualties whatsoever, a fact Scot attributed to God's providence.[65] For their part, the commissioners surveyed the works constantly, arriving earlier and staying later than the laborers themselves. In the meantime, Pett and Baker were occupied in building a sluice of timber, while Digges sent down from London a mason named Symons to design and build a second sluice of stone.[66] On 21 July 1583, Thomas Scott wrote a triumphant letter to Walsingham informing him that the two walls of the pent were finished and seemed fully watertight. The work had been projected to take two years, but a surplus of labor and the favorable weather of that summer made it possible to finish the walls in just over two months, and at a total cost of at least five thousand pounds less than the next lowest estimate.[67]

The decision to commit to the Romney hydraulic construction technique marked a turning point in the administration of the Dover Harbor project. For the first time since the works began, the Privy Council actually deferred to the opinions of the commissioners and advisers they had chosen to develop and execute a plan of action. The councillors certainly did not relinquish their authority in the project; indeed, they expected the commissioners to advise and consult with them at all times, and Thomas Scott was scrupulous in sending his regular reports to London. Yet this time, rather than trying to run the project directly themselves, the privy councillors relied upon the recommendation of their trusted expert adviser, Thomas Digges, and the empirical and administrative experience of Thomas Scott. The brilliant success of the Romney method boosted the credibility and reputation of both men; Digges in particular gained even greater influence with the Privy Council, which in turn increased his status on the Dover commission. He played an active role in all of the commission's subsequent deliberations, in which his opinions virtually always prevailed; and from 1583 onward the Privy Council continued to heed his advice and never again overruled a Dover commission resolution. Digges's unique combination of technical expertise, personal disinterestedness, and political and social connections made him the project's most important expert mediator, the only officer in a position to bring about consensus among all the parties involved and keep the works from stagnating in endless debate.

## The Harbor Mouth Debate: 1584

With the pent seawalls completed, the next major controversy at Dover involved the location of the new harbor's entrance. The issue went far beyond matters of convenience and aesthetics, addressing questions such as the feasibility and safety of ships' entering the harbor under different winds and the efficacy of the new pent and sluice in scouring the harbor's mouth. The Dover commission had already broached the issue, most recently at the beginning of 1583, but had never discussed it at length. Poyntz, while attempting to drain the harbor in 1582, had agreed with the Flemings and opted for an entrance to the east of Thomson's pier; using this location, they felt, would prevent the need to alter or demolish any of Thomson's extant works, saving time and money. Yet by 1584, others had come to believe that opening the mouth through Thomson's western pier might offer other advantages that would justify the extra effort. The issue sparked one of the most contentious debates of the entire project.

The proponents of a western harbor mouth (through Thomson's pier) argued that despite the required demolition, such an entrance would actually be less expensive to construct, since there was a better foundation on which to build the new jetties. The jetties themselves could also be made longer, which would help deflect waves and currents and allow ships to enter the harbor in calmer water and greater safety. Finally, the approach into a western entrance from the open sea would be straighter and more direct, especially given the prevailing southwesterly winds of the area. The advocates of an eastern mouth countered that the difference in cost was actually trivial, given the need to cut through the middle of Thomson's pier. The "firm foundation" for the jetties on the western side would also pose an increased danger to sailors, because it was primarily composed of hard rocks on which errant ships might break up—the eastern mouth, though less direct, was much deeper and had a soft mud base that would pose less of a hazard to ships' hulls. Moreover, though a western entrance might be more direct, it would also leave the inner harbor exposed to the flood tide of the Channel. This would allow waves and "billow" to enter the harbor unchecked, tossing the anchored ships and eventually undermining the foundations for the all-important cross wall and sluice. The two camps were sharply divided, and having no empirical means of settling the issue, they debated largely through speculation and assertion.

The first salvo came in December 1583, in a letter to Walsingham from two of the Dover commissioners, Edward Boys and Henry Palmer. The two claimed to have made a thorough examination of the site and to have consulted with Pett (the carpenter) and Symons (the mason) as well as "with all the master men of any account then in the town." All were agreed, they wrote, that a western harbor mouth was the best option.[68] Their letter is most striking in its paucity of official signatures; despite serving on the commission, Boys and Palmer by themselves held no authority to make decisions in the project, or even to communicate directly with Walsingham. A normal report from the commission might be written by the treasurer (James Hales), the master surveyor (Digges), the chief overseer (Scott), the lieutenant of Dover Castle (Barrey), the warden of the Cinque Ports (Cobham), or anyone, in short, with more authority than Boys and Palmer had. Furthermore, any letter containing more than just a simple progress report was generally written by the commission as a whole and carried the signatures of all members present at the meeting where the letter was drafted. Two commissioners would not even have constituted a quorum and could never have made such a decision on their own. The writing of such a letter to a prominent privy councillor, in fact, implicitly challenged the commission's administrative authority. Theirs, however, was not the last word on the matter.

By May 1584 the location of the harbor entrance was on the (entire) commission's agenda as a matter requiring resolution before the summer's work could begin.[69] Several of the commissioners prepared a lengthy report for the Privy Council, detailing the issues under discussion that spring, though in a highly partisan fashion: the report was divided into three subheadings, "Wherein the east is better than the west," "The discommodities of the west mouth," and "Wherein the one doth resemble the other."[70] Accompanying this (or at least composed at roughly the same time) was a transcription of an interview with "the masters and mariners of Dover, sent for by the mayor of Dover to speak their opinions . . . at the request of the commissioners."[71] The transcript, which included both questions and answers, reiterated most of the information in the commission's report.

The Dover mariners were said to have answered the questions put to them throughout the interview "with one consent . . . no one master there denying any one of them." Interestingly, the second question the interviewers asked them was "[w]hether your opinions were ever heretofore demanded for the convenient making of the mouth," to which the mariners answered that "there was never any such question demanded of them before this present meeting."[72] The commissioners probably included this exchange to undermine the claims of Boys and Palmer that they had spoken "with all the master men of any account then in the town." The transcript of the interview concluded with the mariners' collected signatures and marks, some ten in all. This was officially witnessed by the commissioners and other officers present, who also signed their names: Richard Greynvile, George Cary (two consultants brought to Dover in 1584), William Willis (mayor of Dover), Thomas Wilfford, Barrey, and Digges. Conspicuously missing were Palmer and Boys, as well as many of the more senior commissioners (Hales, Scott, and Cobham).

Thomas Digges was once again a prominent participant in the debate, vigorously advocating an eastern mouth to the Privy Council, both through the commission's report and as an independent adviser. He sent his own twelve-page discourse on the subject directly to the councillors, in which he began by lashing out at his detractors and defending his role in the works, "because I hear it is slanderously informed by some that I cross and hinder the proceeding of the harbor." He recounted at length his many contributions, taking credit for exposing the ignorance and incompetence of those who had opposed him in the past, including Trew, Pett, Borough, and Poyntz. He went on to refute the case for a western mouth point by point and included his estimate of the cost of building such a mouth, amounting to over ten thousand pounds. Finally, he presented his own plan for an eastern mouth and even offered to build it with his own financial backing if necessary, for less than seven thousand pounds. Throughout his report, Digges emphasized his own considerable "experience of all other artificial harbors . . . as well beyond the seas as at home." He insisted that experience was "the most assured teacher" and argued that by means of his experience alone he might "direct myself more certainly, than if I had advice by general consultation with all the mariners and workmen of Christendom."[73] This strategy further emphasized Digges's unique background and value as an expert to the Privy Council; whereas the other Dover commissioners (on both sides of the debate) appealed to the testimony of the Dover mariners, Digges made much of his experience abroad, a claim neither his colleagues nor his rivals could easily match.

In the end, consensus among the Dover commissioners proved unattainable. At a meeting held on 16 June 1584, a week after Digges submitted his own report to the Privy Council, the commissioners and outside consultants were reduced to the rare and inelegant solution of putting the matter to a vote. In a document unique in the archives of the Elizabethan works at Dover, each commissioner and consultant was recorded by name, together with his opinion on the matter. Voting for an eastern mouth were Thomas Digges, Richard Greynvile, George Cary, Richard Barrey, Martin Frobisher, "Mr. Poincts" (most likely Fernando Poyntz, who had not been mentioned in the records of the project since his departure the previous year), Humphrey Bradley (a consultant from the Low Countries), Thomas Scott, Edward Boys (who had weighed in earlier for a western mouth and here changed sides), Thomas Wilfford, and William Partheriche. Voting for a western mouth were William Willis (the mayor, who had previously signed the interview with the mariners advocating an eastern mouth), John Hawkins, William Borough, and Peter Pett.[74] The matter was nominally decided, eleven to four.

Even the vote failed to end to the debate, however. John Hawkins, treasurer of the Royal Navy and one of the minority western-mouth partisans, was apparently so dissatisfied with the outcome of the debate that the commissioners invited him to appeal to Walsingham directly, which he did in writing on the following day. After rehearsing once again his reasons for supporting a western mouth, he stressed that he was not alone in his opinion, claiming that "the townsmen and the masters [of Dover] are of my mind." Together with Pett and Borough, Hawkins also sent a new estimate for a western entrance to Walsingham three days after the vote was taken, apparently at Walsingham's request.[75] But Hawkins's view could not prevail against the united opposition of the commission in Dover and Digges in London. The Privy Council again heeded the advice of its trusted expert, deferring to the large majority of the commissioners, and the eastern mouth was duly constructed over the next two summers.

The debate over the location of the harbor's new entrance is remarkable both for the extraordinary efforts on the part of the project's managers to achieve a consensus of opinion on the matter and for the dismal failure of those efforts. As Mark A. Kishlansky has shown in the context of selecting members of Parliament, consensus was highly valued among the gentry and ruling classes of Elizabethan society, especially when one's betters were watching. Putting the matter to an open vote tended to be seen as a last recourse; it represented an embarrassing failure to find a less publicly contentious, more gentlemanly resolution.[76] In the previous year's controversy over the Romney method, the privy councillors had worked very hard to forge a consensus of opinion among all the parties involved, forcing round after round of deliberations until everyone had either signed on with the Romney plan (e.g., Borough) or left the project altogether (e.g., Poyntz).

The council adopted a similar strategy with the harbor mouth debate; Walsingham in particular went out of his way to hear all sides of the issue, and Burghley reportedly tried to reassure the disputants that "the council doubteth not but that you are all well-willers unto Dover harbor, although there be some difference in your opinions concerning the best place where the haven's mouth shall be."[77] Six days before the commission's contentious vote, on 10 June 1584, the Privy Council even admonished the commissioners to consider "[w]hether upon view of the ground and observation of the tides, the gentlemen that have given their assent to have the mouth open at the east side of the crane [on Thomson's pier] shall find any cause to alter their opinion. And if they shall see cause, then to set down reasons of the said alteration."[78] Nevertheless, the matter was settled only by means of a split vote and was permanently recorded as such. Although the project's history abounded in controversies of a similar nature, this approach to resolution was unique and remains a rare public acknowledgment of irreconcilable differences.

Yet the role of the expert mediator could be most consequential in precisely this sort of administrative crisis. In this case, Thomas Digges was able to help keep the project moving forward in both Dover and London. Throughout the lengthy deliberations, he fought hard at Dover to convince his fellow commissioners to support an eastern harbor mouth. When this seemed unlikely to succeed, his personal report to the Privy Council as their resident expert adviser was important in convincing them to endorse one side, thereby overcoming dissent within the Dover commission and preventing the works from grinding to a halt. In the absence of such trusted expert guidance, the Privy Council might well have allowed the project to become mired in endless conflict and stagnate for an entire summer, as it had during Poyntz's tenure. The expert mediator was thus an important tool not only for achieving consensus but also for coping with a situation in which consensus proved unattainable.

## Epilogue: 1585 and Onward

In 1585, after the new harbor's largest, most important, and most costly structures had been completed, the Dover project fades from the Elizabethan State Papers, though construction continued at a slackened pace for another decade. In 1595 the redesigned harbor was finished at last; it was much bigger than its predecessor, accommodating scores of large ships comfortably, and the backwater pent performed its cleansing function successfully for a number of decades. The Elizabethan version of the harbor has proved to be remarkably long-lived; with some modification, the plan of 1583 reflects the layout of the inner harbor at Dover today.[79] As for the town, the presence of a permanently viable harbor helped to spark an economic boom. The population grew manyfold after 1590, and within fifty years new houses had been built even on top of the seawalls themselves, a testament to their stability and permanence.

The new harbor's ultimate success was transitory, however. The natural forces that condemn Dover to perpetual silting had been pushed back but not eliminated. By 1660 it was clear that the existing pent was not entirely effective; the sluice was placed too far away from the harbor mouth, and by the time the pent waters reached it they had lost much of their scouring force. A new cross wall was built, creating a second pent in front of the first, but over time this too proved insufficient. Attempts to improve the harbor continued throughout the eighteenth and nineteenth centuries until 1871, at which time the present harbor at Dover was completed. The solution to the silting problem was the very same as that proposed by John Thomson in the 1530s, but which he lacked the technological capacity to execute: seawalls and breakwaters were constructed almost three thousand feet into the Channel, forcing most of the sand and mud brought by the tides to be carried so far out to sea that they could not settle in the entrances to the harbor. Even so, the harbor basin itself still requires dredging today.

Of greater interest here than the persistent difficulties of the harbor itself is how the efforts to overcome them can illuminate Elizabethan administrators' strategies

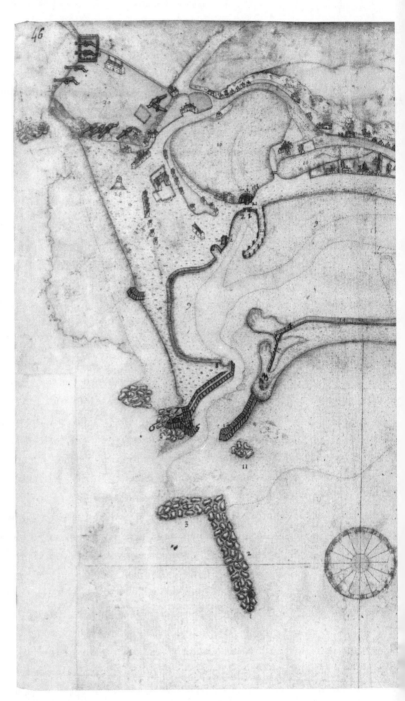

*Figure 2.2.* Plan of the completed harbor at Dover, 1595; north is located at the right-hand side. This map shows how the harbor complex looked in the year during which the final works were added to the harbor mouth. The triangle-shaped pent was located nearest the town, to the right of the plan, with the new harbor itself and the harbor mouth on the left, separated from the pent by the cross wall, which contained the sluice gate. The

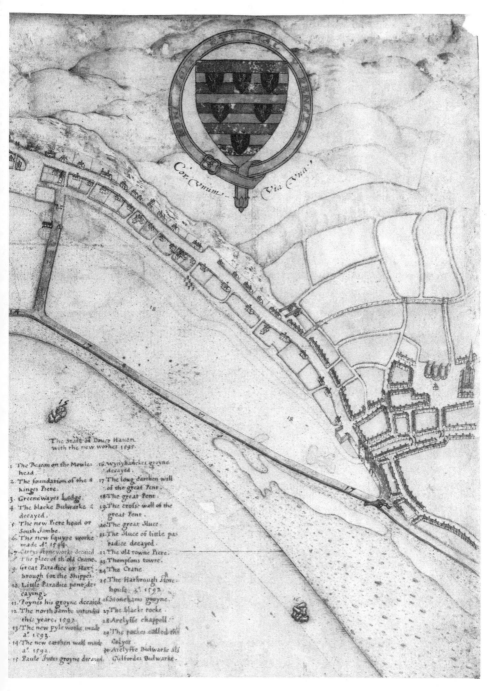

old harbor, a small pool to the upper left of the plan, had been permanently sealed off and was also regulated by a sluice gate. The coat of arms depicted is that of Lord Treasurer William Burghley, and this plan was probably commissioned for presentation to him. Photo used by permission of the British Library, Cotton MSS, Augustus I/i/46.

for tackling complex technical problems. The project to rebuild Dover Harbor during Elizabeth's reign illustrates the high stakes and considerable challenges involved in defining, identifying, and assessing certain types of expertise at a time when there was not yet an established community of experts. Of course, the various administrators of the Dover works were not consciously concerned with anything as abstract as defining expertise. They had a job to be done and were trying to find the best individuals to do it, as successfully and inexpensively as possible. Nevertheless, during such a process, when those who have technical knowledge and practical experience interact directly with administrators who lack it, it is perhaps inevitable that notions of expertise, of precisely what it means to be "the best person for the job," are a primary consideration for the decision makers, whether consciously or not. In effect, whoever controls a project's administration also holds the authority to define *expertise* with respect to the task at hand; those hired are ipso facto identified as the experts. The Elizabethan Privy Council's handling of Dover Harbor is interesting in this regard because the councillors were attempting to shift responsibility for the project from the locality to the center of government. They asserted their authority to make decisions at the local level and brought with them a set of strategies, biases, and priorities that their Kentish counterparts on the Dover commission did not always share.

The increasing power of the Privy Council to oversee local administration throughout the whole of England from their regular meetings in London often involved the councillors in the management of what had traditionally been perceived as local concerns. As a result, they sometimes assumed responsibility for matters they were not fully equipped to handle. For example, in his history of the Dover project, Reginald Scot described Francis Walsingham as a "principal friend to these works" and "the man without whom nothing was done, directing the course, and always looking into the state thereof," a description echoed by many of the Dover commissioners.[80] Yet neither Walsingham nor any of his fellow privy councillors could claim to possess any particular expertise in or qualifications for harbor building. This, indeed, was their reason for seeking assistance and guidance from both local and foreign experts in the first place. Their early missteps in managing the works originated not only in their own lack of expertise, but also in their unwillingness to trust the expert consultants whose services they had obtained.

A comparison with the harbor repairs of Henry VIII's reign is revealing of the Elizabethan councillors' new perception of their own role in the project and their resulting management style. In the 1530s, Dover Harbor was still a local problem, albeit one of importance to the entire realm. When repairs were needed, the initial petitions, the plans agreed upon, and the officers and laborers to execute them all came from the town of Dover. Beyond approving an officer corps chosen from among the locals, Henry VIII's chief minister Thomas Cromwell did little more than receive progress reports and send money; there is no indication that he sought to participate more directly in the administration of the project or felt himself qualified to do so. Despite having a large number of subordinate officers, and despite some backbiting at the works, John Thomson ran things his way, and he viewed the new harbor as his own personal accomplishment.[81] By approaching the king with his own plan and then carrying it out with royal sanction, Thomson *defined himself* as the best man for the job, the only true expert present at Dover Harbor.

Elizabeth's Privy Council, in contrast, played a much more active role. Rather than merely allowing the people of Dover to come up with their own plan, the councillors dispatched consultants and surveyors, men of their own choosing who did not come from Dover, to examine the harbor and report to them directly. They then took it upon themselves to order the creation of a Dover commission to administer the works on their behalf. The members of the commission were locals, but they derived all of their power and authority from the Privy Council, which required them to report their every move to London. Moreover, whenever the commission tried to make a major decision, they were repeatedly overruled from above. It was, after all, the Privy Council's decision to hire John Trew despite the inclinations of its own advisers[82] and to retain Fernando Poyntz's services in the face of stern opposition from the Dover commissioners.

The respective funding schemes for repairing the harbor also indicate a shift in control from Dover to London. During Henry's reign, Thomas Cromwell simply sent money from the royal coffers to Dover on request from the local officers. Despite the king's personal interest in the works, there was relatively little monitoring of the project's finances from London. In the 1580s, however, funding for the project was far more formalized. Money was first raised to begin the works by virtue of letters patent from the queen, granting the Dover commission a monopoly on English corn and beer exports. When the funds from licensing the exports proved inadequate, further money was raised by a parliamentary act of 1581 establishing a new customs fee on the use of London's harbor, and the funds raised by this measure proved more than sufficient. After the London customs fees were collected, they were placed in a special trust administered by the royal customer Thomas Smith, who disbursed them to Dover on the lord treasurer's orders, only after the commission had applied to the Privy Council for more money. The councillors, through their active administrative style and their tight control of the project's purse strings, made it known that the ultimate authority for building a successful harbor at Dover, and by extension the authority to select the best qualified experts for the job, lay solely with them.

Yet by the summer of 1583 this was no longer the case. The debate over the Romney technique for building seawalls proved a turning point, through which the commission regained some local control over the works. The Romney method was heralded by its proponents as a local solution to a local problem; it had been tried and tested with great success in the very same county and under some of the very same administrators; it was presented to the Privy Council by some of the local men most expert in its execution. The councillors were slow to sanction it, insisting upon several rounds of deliberations in London, but in the end the commission's decision won their approval. After the spring of 1583, the Privy Council still expected to be advised of each and every detail of the works and continued to send the occasional consultant or adviser to offer his opinion of the works. But they never again ignored the commission's opinions, imposed their will from above, or overruled a commission decision once it had been made. The commission was allowed a freer hand in the day-to-day management of the works, and by 1584 they were even permitted to petition directly to Thomas Smith for more money as they needed it.[83]

The rapid and unqualified success of the Romney method reassured the reluctant privy councillors and encouraged them to continue to place their trust in those who had proved themselves capable and effective. Thomas Scott's reputation and credibility with the Privy Council rose in 1583, for example, based in large part upon his championing of the method and his skilled oversight of its execution. The councillors came to see in Scott a skilled expert upon whom they could rely to get the job done. Even more important in this regard was Thomas Digges, who was able to base his own authority as an expert mediator not only on his background and experience at Dover and elsewhere, but also on his personal relationships with both the Kentish gentry and the members of the Privy Council. Digges was certainly one of the most technically informed and experienced men involved in the works; but what entitled him to a seat on the Dover commission was his status as a Kentish gentleman, an advantage lacked by other prominent figures such as Wynter, Poyntz, and Borough. In addition, he had worked closely with Walsingham and Burghley as a member of Parliament; this familiarity gave him personal access to the Privy Council and earned him their trust, something the other Dover commissioners could not command so readily. Digges's connections in Dover and London, together with his knowledge of harbor design in the Low Countries, made him an ideal expert mediator and a most influential adviser during the most critical years of the Dover project.

As a key component of Elizabethan centralized administration, the expert mediator was responsible not only for advising and reassuring his patrons and carrying out their orders on site, but also for building consensus among all those involved with a given project. This was especially important at Dover, given the project's several rival claimants to expertise and the contentious debates they spawned. Once again, Thomas Digges played an important administrative role. Digges's status as a prominent member of the Dover commission certainly boosted the commission's credibility in the Privy Council's eyes. Having a trusted expert serving on the commission, and having him available in London to advise them on demand, made it seem less risky for the councillors to take the commission's resolutions more seriously from 1583 onward. Moreover, Digges's obvious influence with the Privy Council probably helped to give his own opinions greater weight with the rest of the commissioners during their deliberations. It was Digges's views, after all, that would receive the most sympathetic hearing in London. As an expert mediator, therefore, Digges was influential in both circles and could do more than any other single figure to bring about the consensus valued by everyone involved.

If Thomas Digges provides an example of expert mediation facilitating successful centralized administration, Thomas Thurland's efforts as a mine manager for the Company of Mines Royal must be seen as a failure of mediation. The two cases do show important similarities, though: both sets of administrators (the company's shareholders and the Privy Council, between whom there was significant overlap) faced substantial technical and managerial challenges, and both turned to expert mediators for help. Those mediators were most effective when they were able to build a consensus among all the parties involved—Thurland's greatest success as a manager and mediator was bringing together the English and the Germans to face threats to their common interests from an outside party, such as Cornelius de Vos. The privy councillors ultimately succeeded at Dover, whereas the company's shareholders failed at Keswick, in large part because the former found an expert whom they trusted. Thurland never acquired the expertise he needed to monitor the copper mining works effectively and so never commanded the trust of his patrons. Digges, in contrast, could move knowledgeably and confidently between the royal court and the harbor construction works; he was thus able to convince his patrons to trust his advice and delegate some of their authority to him and his colleagues on the Dover commission.

The large, expensive, complex projects undertaken by London administrators at Keswick and Dover underscore the growing demand in early modern England not only for various kinds of technical expertise but also for new methods to manage that expertise once it was acquired. The Elizabethan Privy Council and the governors of joint-stock companies in London had a great deal at stake, politically and financially, in the success of these projects, and expert mediators played a pivotal role in making them feasible and profitable. In choosing their experts, the administrators initially had a great deal of discretion in deciding how the relevant expertise would be defined and put to use. As expert mediators became more common and established as tools of sixteenth-century administration, they self-consciously began to define and shape their expertise according to their own ideas. The experts were intended to help bridge the physical and intellectual gap between metropolitan administrators on the one hand and the corps of on-site technical practitioners on the other. Yet the existence of this gap gave the experts who crossed it a lot of leeway in how they portrayed themselves and their expertise to both their patrons and their subordinates.

The following two chapters examine the introduction and development of mathematical navigation in England during the second half of the sixteenth century, as an example of experts seizing control of their field of expertise and redefining it to suit their own agenda. The process of replacing traditional craft practices at sea with new mathematical technologies began when London merchants and investors sought out foreign navigational expertise as a means of restructuring and expanding England's overseas trade. The merchants had started out with a preconceived notion of what constituted contemporary navigational expertise, based in large part upon their perceptions of the Spanish merchant marine and its supporting institutions. By the 1580s, however, English mathematicians had overtaken the process and began to redefine navigational expertise for their patrons in such a way that it sometimes bore little or no resemblance to actual maritime practices of the period. Our story takes us, then, first to the shipmaster's cabin and thence to the mathematician's study.

CHAPTER THREE

# Early Mathematical Navigation in England

> What can be a better or more charitable deed, than to bring them into the way that wander? What can be more difficult than to guide a ship engulfed, where only water and heaven may be seen[?]
> MARTÍN CORTÉS, *The Arte of Nauigation*

> I have known within this 20 years that them that were ancient masters of ships hath derided and mocked them that have occupied their cards and plats, and also the observation of the altitude of the Pole [Star].... Wherefore now judge of their skills, considering that these two points is the principal matters in navigation.
> WILLIAM BOURNE, *A Regiment for the Sea*

On 18 September 1553, Sir Hugh Willoughby anchored his ship in a bay somewhere on the northern coast of Lapland, from which neither he nor his crew would ever sail forth again. His mission was to search for a northeast passage to the lucrative spice markets of Asia, and he commanded a fleet of three English merchant ships for that purpose. Despite his rank of captain general of the small fleet, Willoughby was no mariner but a career military officer who was knighted for his service in Scotland in 1544. He had been chosen for what was to be his final commission in large part because of his reputation as a courageous and capable commander, but his skills in seamanship were sadly lacking, and he had already lost contact with one of his ships in the bitter Arctic weather. Fearing to be at sea when the full harshness of winter fell upon them, he decided to weather the frigid, dark season as best he could with his two remaining ships in whatever haven presented itself. Willoughby's journal of his last days was later found, together with his two ships and the remains of his crew, frozen in the ice of their last anchorage. His final, desperate journal entry, printed subsequently in Richard Hakluyt's *Principall Navigations*, remains a poignant testimony of an early English voyage of exploration gone horribly wrong:

> Thus remaining in this haven the space of a week, seeing the year far spent, and also very evil weather, as frost, snow, and hail, as though it had been the deep of winter, we thought best to winter there. Wherefore we sent out three men south-southwest,

to search if they could find people, who went three days' journey, but could find none. After that, we sent [an]other three westward four days' journey, which also returned without finding any people. Then sent we three men southeast three days' journey, who in like sort returned without finding of people, or any similitude of habitation.

Here endeth Sir Hugh Willoughby his note, which was written with his own hand. [Hakluyt's marginal note][1]

Thirty-four years later, almost to the day, the renowned Arctic explorer John Davis wrote a letter to his patron, the prosperous London merchant William Sanderson, who had undertaken financial responsibility for all three of his voyages (1585–87) to seek a northwest passage to Asia. Davis, a close friend of Humphrey Gilbert and Walter Raleigh, had spent most of his life at sea since boyhood, and by 1587 he was regarded as a highly skilled mariner and "a man very well grounded in the principles of the art of navigation," with a ready command of all the latest mathematical techniques and instruments.[2] Although his 1580s voyages to the northwest were his first commissions of exploration, his future travels would take him all over the world, including several trips to the East Indies. As he wrote his brief letter to Sanderson, Davis was still exhausted, having only just returned to port from his most recent explorations on the previous day. Nevertheless, he wrote in a positive, almost triumphant tone, confident in his abilities as an explorer and navigator despite his failure for the third time in as many years to find the passage he sought:

> Good M. Sanderson, with God's great mercy I have made my safe return in health, with all my company, and have sailed 60 leagues further than my determination at my departure. I have been in 73 degrees [north latitude], finding the sea all open, and 40 leagues between land and land. The passage is most certain, the execution most easy, as at my coming you shall fully know. Yesterday, the 15 of September, I landed all weary, therefore I pray you pardon my shortness. Sandridge, this 16 of September, Anno 1587.
>
>                     Yours equal as mine own, which by trial you shall best know,
>                     John Davis[3]

The art of navigation in England underwent tremendous and expansive changes in the years separating Willoughby's and Davis's respective Arctic voyages. Although English merchantmen had been accustomed to sailing as far afield as Iceland and the Levant during the late fifteenth century, by 1550 few English ships strayed from their (by then) long-familiar trade routes to the staple cloth markets of the Low Countries. Yet by 1570, just twenty years later, English mariners had already undertaken several long-term voyages of Arctic exploration, charting the seas north of Lapland and Russia in search of a northeast passage to Asia and maintaining a regular and profitable trade with Russia via the White Sea. By 1590 they had also made a number of voyages in search of a northwest passage to Asia and had attempted to found permanent colonies in North America. Whereas English explorers initially lost many of their ships and crews to harsh Arctic weather and difficult sailing conditions, over just a few decades they gained a proficiency in Arctic voyaging that bordered on mastery. Of course, English navigation was hardly limited to Arctic waters; English ships also became a common (and fearsome) sight in the Mediterranean and Caribbean seas and along the West African coast. Before the end of the sixteenth century, not one but two Englishmen had taken their ships around the world, a feat that had not been accomplished since Fernando Magellan's voyage of 1519–22. By 1610 English merchant ships had even become a regular presence in the East Indies, albeit not by the northern sailing route they had long sought. How did such striking and rapid advances take place?

Perhaps the most important change in English navigational practice during the latter half of the sixteenth century was the introduction and further development of navigational technologies founded upon mathematics and astronomy. These new technologies included both instruments and techniques—the physical devices themselves together with the set of skills and knowledge required to use them.[4] After all, the astronomical instruments used to make stellar observations were quite useless apart from the calculations in which those observations were subsequently employed. Likewise, the pilot's plane charts were meaningless without a mind-set that could picture the Earth's surface in two dimensions, divided into a geometric grid pattern of latitude and longitude lines—a mind-set that would have been wholly foreign to a traditional, nonmathematical pilot. This dual definition of *technology* is important to understanding English navigational development because, though the acquisition of physical devices was often a simple matter, the adoption of the mental framework in which they had meaning often proved much more difficult. Taken together, however, these new aspects of technology allowed English pilots to sail beyond the shores of northern and western Europe and explore unknown seas.

Before 1550 English pilots had generally steered their way from one familiar port to another using only the traditional methods of North Sea piloting, a centuries-old art learned through long experience at sea, sailing the same well-known routes over and over again. The skills and knowledge acquired by such pilots were therefore fundamentally *local* in nature, based as they were upon direct and repeated personal experience. Pilots became competent in their art with respect to specific places and circumstances only; their knowledge was rooted in space and time and could not readily be transferred to other areas or conditions. A pilot who knew every nuance of the English Channel would have been utterly lost in the Mediterranean Sea, and the deliberate exploration of unknown waters was simply not within his purview.

The introduction of mathematically derived methods of navigation began to change the very idea of what it meant to be a ship's pilot. When applied to the problem of finding one's position on the open sea (in the absence of any fixed landmarks), mathematics had given rise to a host of new solutions, including geometrically based charts and simplified instruments of astronomical observation.[5] Whereas the traditional pilot's skills were at once defined, produced, and limited by his own experience, the mathematical pilot's greatest comparative asset was his ability to navigate with some confidence precisely where his prior experience failed him, and to keep track of his journey as he went. Going beyond the memorization of a series of local conditions and circumstances through constant repetition, mathematics allowed the pilot to begin to cultivate a kind of theoretical knowledge—a set of general principles that could be applied indifferently, wherever he chose to sail. His traditional, local methods could then be expanded into the unknown, through mathematical technologies independent of familiar places or objects. His reliance on familiar, material reference points was thus augmented, if not supplanted, through recourse to abstract mathematical data.

Yet although the fruitful connections between mathematics and maritime exploration might seem natural to modern eyes, the future preeminence of mathematical navigation must have been far from obvious to the average English pilot of 1550. The practical value of such novel technologies cannot have been immediately apparent, after all, to mariners who had been successfully guiding their ships to their intended ports without them by centuries-old methods. The new mathematical techniques were often difficult and tedious to learn and required some expensive instruments whose maritime utility was yet unproven, from that average pilot's perspective. If their traditional methods were by and large sufficient for their needs, why should the pilots bother to acquire the new technologies? Moreover, the art of piloting tended to be highly conservative in nature, given that a new pilot's training rarely moved far beyond what his own master pilot might teach him. It took many years of on-the-job training to become a pilot, and a boy's life at sea did not allow much opportunity for more formal education. How was it, then, that mathematics became such a prominent feature of English navigational practice in such a short span of time?[6]

The successful introduction of mathematical navigation technologies to the

English maritime context required that two key conditions be met. First, English pilots needed a compelling reason to innovate, and in this case the motivation came not from the pilots themselves, but from the merchants and investors who employed them. By the middle of the sixteenth century, the precipitous decline of the vital English cloth trade in the Low Countries, together with a general perception that the English were falling dangerously behind their Iberian rivals, had convinced a number of prominent merchants of the need to acquire the latest navigational methods. This, they hoped, would help them to scout out and exploit new and more lucrative overseas markets, especially the importation of East Indian spices and other valuable Asian commodities. Second, the English pilots themselves needed innovators and teachers, officers of recognized authority who could introduce the new technologies and impose them upon a community of conservative and potentially resistant practitioners. In England this role was filled by Sebastian Cabot and Stephen Borough, the two masters most responsible for training the succeeding generation of English navigators.

Cabot and Borough, however, did not simply base their authority as master pilots on their command of mathematical navigation per se, the value of which still remained to be proved in most English seamen's eyes. Rather, they were known and respected as skilled navigators beforehand, in part because of their reputations as experienced maritime explorers and in part through their mutual association with an institution of internationally recognized navigational authority, the Spanish Casa de Contratación in Seville. While Cabot's and Borough's long experience at sea lent them credibility among practicing mariners, their connections to the renowned Casa further impressed the wealthy and powerful English merchants who hoped to duplicate Spain's success in building a global trading empire. Their dual authority with patrons and practitioners, derived from their command of both the theory and the practice of mathematical navigation, made Cabot and Borough ideal mediators of navigational expertise. Each possessed the practical skill necessary to train and work with practicing English pilots, as well as a knowledge of the more abstract and systematic navigational theory taught at the Casa, which so impressed English investors. With the patrons' power behind them, Cabot and Borough used their expertise to gain considerable control over navigational training in England and were thus able to redefine the navigator's art according to their own ideals.

It can be difficult to hear the voices of practicing English pilots, especially when attempting to understand what they were actually doing at sea. Most sixteenth-century mariners wrote no treatises on their art and left few artifactual remains. Whatever books and instruments they took to sea with them are nearly all lost to us. A trace of them infrequently turns up in company records or ships' inventories, and a few instruments have been recovered from contemporary shipwrecks, but such examples are frustratingly rare. Moreover, it is likely that of the small number of charts and instruments that do survive, most were never actually taken to sea but were owned instead by land-bound collectors, patrons with an interest in maritime activities. Apart from these few artifacts, we have only the narrative accounts that the mariners themselves wrote of their voyages—logs, journals, letters—and scattered documents relating to their preparation for voyages of exploration. These materials are relatively numerous and easily accessible (thanks mostly to the efforts of the younger Richard Hakluyt, who in the late sixteenth century compiled, edited, translated, and published a large collection of original documents pertaining to English exploration in his *Principall Navigations*), but they also vary considerably in quality and degree of detail. Some journals were kept by the ship's master himself, for example, but many more were written by less navigationally minded officers, and references to actual navigational practices (as opposed to mere data) can be scarce. The historian must also be careful not to place too much emphasis on the voyages for which such documents survive. The details of long-term voyages of exploration were invariably better recorded than those of humbler voyages to European ports, for example, yet the latter were far more numerous and constituted the sole occupation of the vast majority of English pilots and ships. Focusing too strongly on the explorers at the expense of more mainstream pilots could thus give a faulty impression of common English practices at sea. Nevertheless, if cautiously used, these documents can help us begin to determine whether and when certain mathematical technologies of navigation were actually employed at sea, how often, and with what success.

## The Art of Piloting, Old and New

Until the middle of the sixteenth century, the skills of an English ship's pilot were profoundly local in nature, founded upon a thorough personal knowledge of the coasts and waters he sailed.[7] Pilots learned their craft only through many years of experience at sea, under the supervision of an older, more established pilot, sometimes within the context of a formal apprenticeship. The good pilot was always a highly experienced one, who knew exactly where all the undersea dangers were located along his routes (or where they were likely to have shifted to) and had memorized the safe passages through or around them. He could recognize the correct approach to different ports of call by means of local landmarks, such as church steeples or prominent trees, which he used as fixed reference points to guide him through any hazards not directly visible to him. He had considerable firsthand knowledge of the tidal patterns of all the areas he sailed, their timing and range, as well as the direction and intensity of the local currents and prevailing winds. Given the strong tides and tricky sands of the Thames estuary and the English Channel, plus the shallowness of many Channel harbors, a pilot's local knowledge was (and is) essential, the difference between safely reaching and entering a port and running aground on the way there. The traditional art of piloting still survives, though only under the most local of circumstances: today, busy ports employ locally trained master harbor pilots, who assume control of any and all ships seeking to enter, guiding them safely into anchorage where the ship's own navigator might founder.[8]

In sailing from one port to another, the simplest course for a sixteenth-century pilot to follow might be to hug the coastline all the way, watching for familiar landmarks to confirm his position as he went. For this the pilot would need nothing other than a thorough command of a host of local navigational details, especially concerning the local tides, winds, and currents, the underwater hazards he might expect to encounter, and the landmarks he could use to help steer around them. This simple-sounding tactic was inefficient, difficult, and dangerous in practice, however. Besides making his journey considerably longer than a more direct route would, staying within sight of a winding coastline could prove very challenging in a sailing vessel, if the winds did not cooperate. Hugging the coast also maximized the chances of encountering sands and shoals in shallow waters or running aground in foul weather. For these reasons, few if any sixteenth-century pilots are likely to have used this method. Following the coastline would have been of very limited use to English shipping in any case; the pilot of any merchantman intending to trade overseas would obviously have to sail in open waters at some point and therefore navigate without recourse to landmarks.

But even when sailing in open waters, English pilots relied heavily upon their local knowledge and needed few other tools. In 1550 their principal means of navigation was usually dead reckoning, for which they used only a magnetic compass, a recording device called a traverse board, and a compilation of navigational data called a rutter. The typical rutter was little more than a list of sailing directions, to be used in setting courses: the compass headings to be followed between one known point and another, estimates of the distances between them, information regarding tides and water depth, and perhaps some sketches of coastal landmarks to be recognized along the way.[9] The pilot steered his ship along the headings recommended in his rutter, as consistently as the winds and currents would allow, using his magnetic compass as a guide. As he sailed, he kept track of the amount of time each heading was followed, using the ship's standard half-hour watch glass, and recorded the information using the traverse board; he also estimated the best he could the ship's speed during each time interval. Using the ship's presumed speed and duration along a given heading, he could then determine how far the ship had traveled in which direction and estimate the time of his expected landfall accordingly. Although Mediterranean pilots also used portolan or plane charts to help them plot their courses, charts were seldom used among North Sea navigators before 1550.

Dead reckoning could be a dependable method of navigation, but its use involved a host of uncertainties. The rutters' navigational data were often inaccurate, and even when they were not, the pilot's assessment of his ship's speed and heading were both little more than educated guesses. Sailing ships were at the mercy of winds and currents that pushed them away from their intended course, and although sixteenth-century vessels could sail into the wind by tacking, this greatly added to the complexity and uncertainty of the pilot's estimation of his progress. Gauging the speed of one's ship was an especially imprecise art, more a matter of feel than of objective measurement. To make matters worse, the basic multiplication of speed by time to get the distance traveled was not necessarily an easy calculation for ships' pilots, most of whom had little or no formal education and whose mathematical skills were often rudimentary. Even the magnetic compass, a medieval invention, could be a confusing and misleading device outside familiar waters; in addition to the fact that a great many were imperfectly made, during the first half of the sixteenth century, no systematic attempt was made to take the Earth's magnetic variation into account. Magnetic north could therefore shift unpredictably, leaving the pilot with no way of knowing whether his course was true. Only a pilot's thorough knowledge of the details, tendencies, and idiosyncrasies of both his ship and his route, gained through endless repetition of prior experience, could help him to compensate for these uncertainties.

In the comparatively shallow waters of the North Sea, the pilot did have another means of determining his position: he could use his lead and line to measure the water's depth and check the condition of the seabed beneath his ship. The lead and line consisted of a large lead bob tied to the end of a long cord, with distance units marked by colored ribbons tied at fixed intervals. With the ship at rest, the pilot dropped the weight into the water and kept track of the length of cord extended until the weight reached the bottom, in order to find the water's depth. Moreover, having coated the bottom of the weight with tallow, he could then bring up a small sample of the seabed, to determine its composition (white sand, brown pebbles, etc.). A highly experienced pilot could use these two pieces of informa-

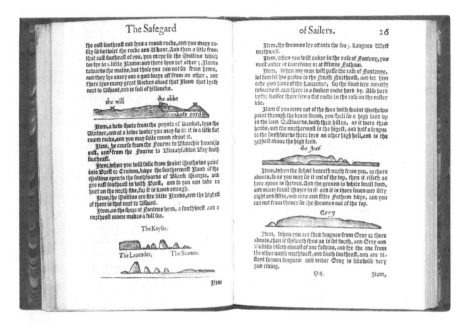

*Figure 3.1.* A typical page (fol. 25 v–26 r) from Cornelis Antoniszoon, *The Safeguard of Sailers: or great Rutter* . . . , trans. Robert Norman, 2d ed. (London, 1587). Although this example is taken from a popular printed text, most rutters circulated in manuscript. The work contained little more than a collection of compass headings, estimated distances, and rough sketches of coastal features and landmarks by which the pilot was supposed to guide his ship. It focused especially on the problem of negotiating treacherous coastal shallows and harbor entries. Photo courtesy of the Beinecke Rare Book and Manuscript Library.

tion to figure out roughly where his ship was, relative to his intended destination. In essence, he was able to use the lead and line to navigate using what may be thought of as an underwater frame of reference. A rutter for the North Sea might even contain information regarding bottom depths and conditions as a means of confirming one's dead-reckoning course.

The pilot's knowledge, then, was specific and fundamentally local in nature. His art was rooted in his long personal experience in sailing familiar waters; the more often he had sailed a given route, the better qualified he was as a pilot *for that particular route*. A good pilot knew roughly how long it took to sail from one port to another in all types of weather, what winds and currents would help or hinder him along the way, where all the underwater hazards were, and the precise bearing to maintain when in sight of a particular church steeple, because he had seen and done it dozens if not hundreds of times before. Prior experience was his only reliable guide. Although his rutter probably included data telling him how to sail by dead

*Figure 3.2.* A basic traverse board. The instrument served as a mnemonic device, on which the pilot recorded with pegs both the heading followed by the ship during each turn of the watch glass (a half hour in duration) and the estimated speed of the ship during that time. These data could then be used to determine the ship's dead-reckoning position. This particular example has been tentatively identified as Scandinavian, c. 1800, but the design would have changed little from the late sixteenth century onward. Photo courtesy of the National Maritime Museum, London.

reckoning to an unfamiliar port, that information was nearly useless by itself, and only a reckless pilot would entrust the safety of his ship to it; far better to find and commission the services of a pilot who had the required experience. The traditional pilot's art, in short, was not portable. An English pilot in 1550 was almost as tightly rooted to his familiar routes as the hilltop trees by which he might guide his ship into port.

For many decades, this did not present a problem. During the first half of the sixteenth century, as English trade became ever more narrowly focused upon a few staple market cities, English merchantmen sailed almost exclusively to ports in the Low Countries, northern and western France, and Iberia. English pilots rarely guided their ships farther afield than Seville or Cadiz in southern Spain and had no compelling need to wander from their long-accustomed routes. They traveled only sporadically to the Mediterranean, rarely entered the Baltic, and did not cross the Atlantic. Whatever exotic luxury goods from distant markets were sold in England arrived there either in foreign ships or via the staple markets in the Low Countries, especially Antwerp. The fifteenth-century English trade with Iceland, Norway, and the Baltic Sea ports had sharply declined, and the North American voyages of exploration undertaken by the Venetian explorer John Cabot for Henry VII in 1497 and 1498 were not followed up by their Bristol investors. Although a few English fishermen still ventured as far as Iceland or even Newfoundland for cod, they did so in the company of their Continental cohorts and may be seen as an exception to the rule; in any case, they did not record their routes, perhaps to prevent unwanted competition.[10] Even the Royal Navy, greatly augmented by Henry VIII, served mostly as an instrument of coastal defense and was generally confined to the Channel waters. English pilots of the early sixteenth century thus had little need for navigational innovations, which might only tell them how to navigate in places where they did not need to go.

The strained economic circumstances of the 1540s and 1550s altered the situation. The market for English wool at Antwerp, the principal staple outlet for what was by far England's most important export, became saturated and fell into decline. Yet the English, who had become dangerously dependent upon a single export commodity, continued to overproduce wool.[11] The rapid debasement of English coinage during the same period, in order to finance ill-fated wars with France and Scotland, made the English merchants' bargaining position on the Continent even worse, as did the religious turmoil created by England's separation from the Roman Catholic Church. By the time of Henry VIII's death in 1547, English trade was in a state of crisis, and a few influential members of the Privy Council of the young King Edward VI realized that new strategies and markets for English wool would have to be sought quickly. Meanwhile, the dissolution of the English monasteries and the secular redistribution of their land and wealth, together with the (involuntary) liquidation of assets once tied up in the wool trade, had created a large amount of mobile capital in England, which many merchants and courtiers were willing to invest in new overseas trading ventures.[12]

East Indian spices were among the most lucrative of all the commodities traded in sixteenth-century Europe, and English merchants were increasingly eager for the opportunity to peddle them. But there were only two known maritime routes from Europe to the East Indies: the Straits of Magellan, controlled and protected by the Spanish; and the Cape of Good Hope, jealously guarded by the Portuguese. To exploit those trade routes, the English would have to battle the entrenched opposition of the much larger, wealthier, and more sophisticated Iberian trading empires. In theory, however, the two southern passages were not the only routes available. As early as 1497, John Cabot had considered the possibility of a northern passage to the Pacific Ocean, and Robert Thorne, an English merchant living in Seville, had tried to persuade Henry VIII to send out an exploratory voyage to the north in 1527, but their suggestions had not been seriously pursued.[13] By midcentury, however, several London-based English merchants had become anxious supporters of overseas exploration. With the encouragement and assistance of John Dudley, Duke of Northumberland and lord president of Edward VI's Privy Council during the king's minority, the merchants determined to find a northern route to Asia and began seeking a navigator with the required expertise for the job.

Exploratory navigation was by definition an undertaking very different from traditional piloting. The whole point, after all, was for the pilot to travel *beyond* the limits of his prior experience and sail in unknown waters, recording the navigational details of his journey along the way so that it might be safely repeated by others. The explorer thus had to leave behind his vast local knowledge, his greatest strength, and find his way by another means. This posed a serious problem; "What can be more difficult," wondered the Spanish cosmographer Martín Cortés, "than to guide a ship engulfed, where only water and heaven may be seen[?]"[14] The solution was to make use of the only fixed and unchanging points of reference available on the open sea: the stars. During the fifteenth and sixteenth centuries, Portuguese explorers had sailed down the West African coast and into the Indian Ocean, while the Spanish explored the Caribbean, down the Atlantic coast of South America, and into the Pacific. Whenever they pushed beyond familiar waters into unknown seas, the Iberians' vast body of local knowledge could no longer help them. They responded by turning to the heavens, incorporating new navigational techniques derived from mathematics and astronomy, and thereby extending the traditional art of piloting into new and uncertain circumstances.

Looking to the skies for navigational clues was not new, of course—European pilots had used the stars to find their heading for centuries before the magnetic compass was invented. But the Iberians learned to put the stars to a new use: finding their north-south position on the Earth, with respect to their points of origin and destination. A pilot could observe the angular altitude of the pole star above the horizon to determine how far north or south he had already traveled and how far he had yet to go. Pilots measured the star's altitude using either a quadrant, a cross-staff, or a mariner's astrolabe (a greatly scaled-down version of its traditional and more complicated astronomical cousin). Like the lead and line measurements of North Sea pilots, the stars thus provided a valuable check against which one's dead-reckoning calculations might be compared. Yet whereas the lead and line depended upon the pilot's intimate familiarity with the seabed over which he sailed, the stars could be used even in unfamiliar waters.

Initially even these astronomical data were strictly local and relative; pilots compared the measured altitude of the pole star at their present location with the known value for their destination and plotted their remaining course accordingly. Over time, however, the measurements acquired a more abstract meaning; pilots gradually came to see the north-south position on the globe as an absolute measurement in itself, a number rather than a comparison, and so learned to think in terms of their *latitude*. As the Portuguese sailed ever farther to the south, the pole star sank ever lower in the sky, and it finally disappeared below the horizon as they crossed the Equator. Needing a new method that could be used in southern waters, they developed a sophisticated technique for calculating their latitude using the altitude of the sun at noon and its location along the ecliptic (which depended upon the time of year). This new technique, though more complicated, could be applied anywhere that an accurate solar altitude might be measured. Crucially, it could be used not only to confirm one's dead-reckoning position in known waters but also to record the latitude of any newly discovered territories, so that one might find them again more easily in the future.

As a result of their extensive explorations, by 1550 Iberian pilots were supposed to be adept at taking altitude measurements of the sun and fixed stars, and every licensed pilot owned the basic astronomical instruments and tables needed to make the required observations and calculations. Even if relatively few pilots had the education necessary to understand the mathematical principles behind their new methods, they had ceased to think only in terms of linear distance and direction (as was required for dead reckoning) and had begun to perceive the Earth's surface

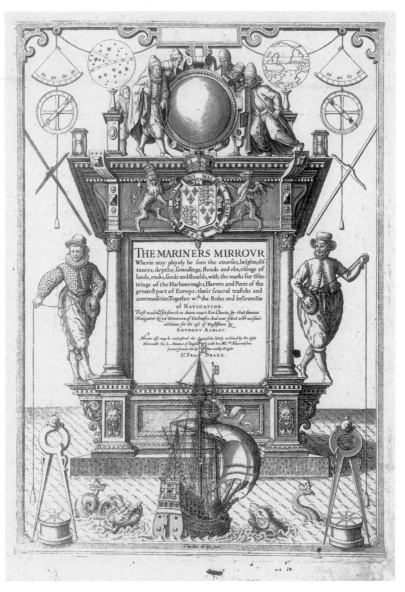

*Figure 3.3.* The title page from Lucas Jansz Wagenaer, *Mariners Mirrovr*, trans. Anthony Ashley (London, [1588?]). It depicts a number of navigational instruments, including the quadrant, the astrolabe, the cross-staff, and the compass. Each of the male figures flanking the title in the center is holding a lead and line. The design of this title page is based closely on the original in Wagenaer's *Spieghel der Zeevaerdt* (Leiden: 1584–85) and was used in a number of editions and translations of the book. This English example, however, has been reworked to appeal to a more partisan audience; the costumes of the flanking figures reflect sixteenth-century English dress, and the cross of St. George flies from the topmast of the ship at the bottom of the page. Photo courtesy of the John Carter Brown Library at Brown University.

*Figure 3.4.* Using a cross-staff to record the altitude of a celestial object. The device consisted of two interlocking crosspieces, the smaller of which slid back and forth along the larger. The navigator held the end of the staff as close to his eye as possible and slid the crosspiece until one tip of it appeared to touch the horizon, with the other tip on the object to be observed. Holding the crosspiece in place, he then read the object's altitude off the longer, calibrated piece. The cross-staff worked best for measuring the altitudes of lower objects, less than 45 degrees above the horizon, because the navigator would have trouble winking his eye back and forth over a larger angle. It was also problematic for measuring solar altitudes, because of the necessity of gazing at the sun. Pedro de Medina, *Regimie[n]to de nauegacio[n]* . . . (Seville, 1563), fol. xxxv v. Photo courtesy of the Beinecke Rare Book and Manuscript Library.

*Figure 3.5.* Using a mariner's astrolabe to record the altitude of the sun. Suspending the astrolabe so that it hung perpendicular to the ground (or deck), the navigator allowed the sun to shine through a pinhole in one of the sights mounted on the alidade. He then lined up the point of light on the hole of the opposite sight by rotating the alidade. When the two holes were thus aligned, the navigator read the sun's altitude from the calibrated circumference of the instrument. The astrolabe was most useful where the cross-staff was weakest: in measuring altitudes above 45°, and especially solar altitudes. It could be difficult to use aboard ship, however, since any breeze could set it swinging or spinning; most mariners' astrolabes had open interiors and were heavily weighted at the bottom in order to resist the wind better. Pedro de Medina, *Regimie[n]to de nauegacio[n]* . . . (Seville, 1563), fol. xv v. Photo courtesy of the Beinecke Rare Book and Manuscript Library.

geometrically, in terms of angular distances on a sphere. Accordingly, the traditional portolan charts long used in the Mediterranean Sea began to include not only the newly discovered coasts but latitude scales along the charts' margins as well.[15]

Mathematical navigation represented a significant advance over traditional piloting; it supplemented pilots' local knowledge and skill with a body of abstracted navigational theory, giving pilots a frame of reference even where they had no prior experience. Using the new astronomically based technologies, Iberian pilots could guide their ships into unknown seas and still have some idea of where they were and how to get home. Once home, moreover, they could provide more useful information about how to return to where they had been. The new technologies were more systematic and general in their potential application, portable in a way that traditional piloting was not, transforming the pilots who mastered them from local craft practitioners into theoretically trained navigators. This abstraction and expansion of navigational knowledge, from the particular, local, and empirical to the general, universal, and theoretical, represented a watershed in the development of navigational expertise. No longer could a pilot possessing only a fixed body of local knowledge be considered a navigational expert, compared with one who had also mastered a more systematic approach that could be applied virtually anywhere.

Before 1550, English pilots did not share their Iberian contemporaries' need for the new navigational technologies, because while Iberian trade had expanded through exploration, English trade had contracted to just a few key European ports and routes. Most English mariners, conservative in their training, neither needed nor wanted to acquire the new mathematically based techniques. Indeed, they would have found it very difficult to do so in any case; the midcentury English maritime community lacked the sort of formal institutions, such as the Casa de Contratación, that had facilitated the training of Iberian pilots in the new methods, and the Spanish and Portuguese were hardly eager to share the secrets of their success with potential competitors. The quest to introduce mathematical navigation to England began not with English pilots but with the merchants and investors who hired them, as they sought to recover and expand upon their former profits through maritime exploration.

## Sebastian Cabot and the First Northeast Ventures

The London-based merchants and courtiers seeking to foster English exploration had a great respect for Spanish success in building a global trading empire and believed that the navigational training provided at the Casa de Contratación

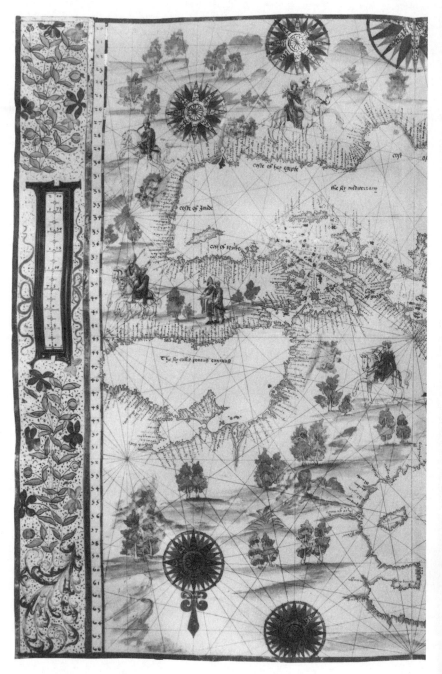

*Figure 3.6.* A plane, or portolan chart of the Mediterranean Sea; north is located at the bottom. The latitude scales on the left- and right-hand sides of this chart do not agree. This was most likely an artifact of the magnetic variation of the compass across the Mediterranean region; although a poor corrective, it was a common (and much criticized)

feature of sixteenth-century charts. This particular chart was certainly never taken to sea; it is contained within an illuminated manuscript atlas presented to King Henry VIII (Jean Rotz, "Boke of Idrography," 1542). Photo used by permission of the British Library, Royal MSS, 20.E.IX, fol. 20.

in Seville had played a major role in the process. Hoping to emulate Spanish achievements, therefore, they sought to acquire the requisite expertise at its source. In 1547, shortly after Edward VI's accession to the throne, members of the new Privy Council (almost certainly led by John Dudley) made overtures to the pilot major of Spain, Sebastian Cabot, to lure him into English service. The second son of John Cabot, Sebastian had left England for Spain in 1512, frustrated with the half-hearted support he had received from the young King Henry VIII after the death of Henry VII in 1509. In 1547 he seemed to the privy councillors an ideal candidate to jump-start England's search for new overseas markets. As pilot major at the Casa, he was responsible for the training and certification of every pilot in the Spanish navy and merchant marine; he understood the mathematical technologies at the cutting edge of Iberian navigational practice and oversaw the construction of the relevant instruments; he had considerable knowledge of the organizational infrastructure through which Spain supported its global trading empire; he was, or at least represented himself to be, an experienced mariner and officer in his own right, who had commanded multiple voyages of exploration for both England and Spain; and by the 1540s, he was even claiming to be an English subject, born in Bristol during his father's service to Henry VII, though this claim was almost certainly fallacious. Cabot, then, appeared to unite in his own person the new mathematical theory of navigation and its actual application at sea, a combination his English suitors believed to be at the heart of Spanish maritime success.[16]

Sebastian Cabot left Seville in 1548 and settled in Bristol; he never left England again, despite repeated Spanish attempts to woo him back to his post at the Casa. He received a generous stipend from Edward VI and was soon put to work for it; within two years, he had put together a modest mercantile training mission of sorts. On 13 November 1550, the *Barke Aucher* departed Gravesend Harbor near London for the Levant ports of Candia (Crete) and Chios, under the command of Roger Bodenham, an experienced Bristol mariner handpicked by Cabot for the job. Though small in scale, the voyage gave the English mariners and merchants involved in it exactly the type of practical experience they so badly needed, sailing and trading in waters that were not familiar to them. The *Aucher* ran into some trouble from the Ottoman Turks, who were preparing their navy for action in the area, but nevertheless managed to deliver their cargo and return home safely, with a respectable profit. Beyond the mercantile achievement, moreover, the voyage's pedagogical mission was also very profitable; Bodenham later reported that "all those mariners that were in my said ship, which were, besides boys, three-score and ten, for the most part were within five or six years after, able to take charge [of ships], and did." Among those serving in Bodenham's crew were Matthew Baker, a ship's carpenter who later became Queen Elizabeth's master shipbuilder, and Richard Chancellor, the chief pilot of Sir Hugh Willoughby's ill-fated voyage of 1553 and master of the only ship in Willoughby's fleet to make it home to London.[17]

Back in England, meanwhile, Cabot played an important role in arranging financial support for a major exploratory voyage to the northeast. Overseas exploration was an expensive and risky undertaking; because no single English investor could afford to hazard the sums required to launch such a venture, Cabot convinced a group of London merchants to share the risk and invest collectively in what became the first of several English joint-stock trading companies. Upon its subsequent incorporation in 1555, the organization was formally called the "Merchants adventurers of England, for the discovery of lands, territories, isles, dominions, and seigniories unknown, and not before that late adventure or enterprise by sea or navigation, commonly frequented," but it came to be known simply as the Muscovy Company.[18] The company's full name shows that exploration and discovery were every bit as integral to its mission as the trade that would predictably result, and the Crown soon granted the investors a monopoly on all exploration to be undertaken "northwards, northeastwards, or northwestwards . . . by sea."[19] The company had not only the Privy Council's blessing but quite a lot of its money as well; although the enterprise was not officially a royal undertaking, many prominent courtiers and privy councillors were among the original shareholders. In recognition of his leadership in organizing the venture, Cabot was appointed as the company's first governor, a post he held until his death.[20]

English mathematicians and mathematically minded navigators were very active in preparing for the Muscovy Company's maiden voyage, sent out in search of a northeast passage to the East Indies in 1553, the first documented English voyage of exploration to the north since John Cabot's in the late fifteenth century. Most prominent among the organizers, of course, was Sebastian Cabot himself. Besides his role in arranging the funding for the voyage through the organization of the Muscovy Company, Cabot very likely provided the original impetus for exploring the northeastern seas, given that his father had hoped to find a passage that way decades earlier. Richard Eden, a contemporary translator of various geographic treatises into English, wrote that Cabot had "long before had this secret in his mind."[21] Cabot had already provided the company's mariners and merchants with the best hands-on training he could give them, in organizing Bodenham's 1551 voyage to the Levant. Finally, in the last stages of preparation in 1553, Cabot compiled a set of instructions for the mariners and merchants to take with them on their mission. The focus of the instructions is divided, with a large number specifically devoted to matters of trade and commerce, which was after all the main purpose of the trip.[22] Of those addressing issues of navigation and shipboard governance, their often rudimentary character illustrates not only Cabot's pedagogical concerns and priorities, but also the grave inexperience of the English mariners for whom he wrote.

Cabot's very first instruction reflected the restructuring of traditional maritime governance among English merchantmen, made necessary by the novel collaborative nature of the Muscovy Company. Before the introduction of the joint-stock system of finance, nearly all English merchants sent out their own ship or ships, trading under their own individual auspices and at their own risk. Joint ventures involving a fleet of ships, such as the Muscovy Company's first exploratory voyage, were unusual and required an unaccustomed degree of coordination and cooperation from all interested parties. Cabot therefore decreed from the outset that "the captain general, with the pilot major, the masters, merchants, and other officers" of the fleet were to work closely together, bound "in unity, love, conformity, and obedience in every degree on all sides," in order that "no dissension, variance, or contention may rise or spring betwixt them and the mariners of this company, to the damage or hindrance of the voyage."[23] By requiring all the masters and pilots of the fleet to work together as a team, Cabot underscored the extraordinary challenges of the Muscovy Company's planned undertaking, as well as the need for the inexperienced English explorers to support one another in such a hazardous endeavor.

The next four points of Cabot's instructions all served to reinforce his first edict by reminding each mariner of his oath to obey the captain general in all things, "with the assent of the council and assistants . . . during the whole navigation and voyage."[24] Although obedience to the captain and the ships' masters was of course customary, the naming of a council to assist the captain was not. Instead of vesting authority for the voyage only in the fleet's commanding officers (the captain general and the pilot major), Cabot stipulated that a council of twelve should be involved in all the major decisions of the mission; these twelve were to consist of the captain general, the pilot major, the three ships' masters and their mates, the minister, a gentleman, and two merchants.[25] This peculiar command structure again reflected the extraordinary degree of cooperation required for such an unusual voyage, and it may also have indicated Cabot's concerns regarding the inexperience and unproven seamanship of his officers. His fifth point of instruction even extended the command-by-council system to the navigational responsibilities for the voyage, specifically naming the several officers who were to work together in setting and maintaining the fleet's course. Not only did "the captain, pilot-major, masters, and masters' mates" each have a say in the process, but no decision could be finalized until "the most number" of the entire command-council approved it, although the captain general was allowed "a double voice" in all deliberations.[26]

This was a striking arrangement; sixteenth-century English pilots of merchantmen were not accustomed to navigating by committee. Very few such ships even had a distinct captain, almost always sailing under the command and guidance of a single master, with a mate to assist him. The presence of both a captain (sometimes the ship's owner) and a master on the same ship already created a situation of overlapping authority. Navigation by such a large committee, even with the commander having "a double voice," was unheard of. It was not common to the Spanish practice with which Cabot was most familiar, either. Though Spanish maritime authorities required their transatlantic shipping to travel in convoys from the 1540s onward, requiring a high degree of coordination among the pilots involved, they apparently only met together as a committee in unusual circumstances, normally following the lead of the fleet's pilot major.[27] On the Muscovy Company's maiden voyage, however, there may well have been no single officer whom Cabot considered competent and experienced enough to assume full control of the entire fleet. In later voyages, by contrast, consultation with other officers was sometimes required, but the authority to make decisions was vested in only a very few.[28]

In addition to safeguarding the fleet, Cabot was also concerned with recording its navigational data as extensively and accurately as possible for future use, and a collaborative effort would have been more likely to achieve this than a single inexperienced and unsupervised pilot. Cabot's seventh instruction specifically addressed the navigational records to be kept throughout the fleet's journey. The "merchants, and other skillful persons in writing," he ordered, were to "daily write, describe, and put in memory the navigation of every day and night" independently—indicating that some of the fleet's officers were not sufficiently literate to do so and would therefore need the scribal help of the merchants on board. The masters of the ships would then assemble once a week and compare their records, "to the intent it may appear wherein the notes do agree, and wherein they dissent." Then, with "good debatement, deliberation, and conclusion determined," the masters were to compile a final, accurate navigational journal for their entire voyage and "put the same into a common ledger, to remain of record for the company." The ships' masters were responsible not only for recording the traditional elements of their craft—"the points [of the compass], and observation of the lands, tides, [and] elements"—but were also expected to have mastered the new mathematical technologies, such as making regular observations of the "altitude of the sun, [and the] course of the moon and stars." Toward this end, Cabot called for "the like order to be kept in proportioning of the cards [charts], astrolabes, and other instruments prepared for the voyage, at the charge of the company."[29]

The reason for such careful attention to the documentation of the fleet's navigational progress was that without an accurate (and duplicable) record of the voyage, nothing the fleet accomplished in its explorations could be translated into a sustained and profitable trade. The main goal of this first voyage, therefore, was not merely to establish an English presence in the spice markets of Asia but also to compile all of the fleet's navigational data into a useful new rutter for journeys to the northeast, so that the company's pilots could guide their ships back to those markets in the future. Again, Cabot believed that this could be most reliably accomplished through a collaborative process of data collection and comparison, and in this his instructions reflected the approach of the cosmographers of Spain's Casa de Contratación, where he had served for so many years as pilot major.[30] The utter lack of experience on the part of his English pilots, who were unused to the technologies of mathematical navigation and had never traveled across the seas in question before, further justified the cooperative approach.

The items missing from Cabot's guidelines are as interesting as those he included. In particular, Cabot provided no written instruction whatsoever on how to use any of the new navigational equipment included for the voyage, in contrast to the myriad maritime instructional manuals printed before the end of the sixteenth century, in England and elsewhere. For Cabot, piloting a ship remained an art learned not by reading (which some of his officers apparently could not do in any case) but through personal observation and practice, as on Bodenham's Mediterranean voyage of 1551. Cabot's commitment to learning through experience was stressed in his fifteenth instruction, in which he ordered that the younger members of the fleet's crew were "to be brought up according to the laudable order and use of the sea, as well in learning of navigation, as in exercising of that which to them appertaineth."[31]

Cabot also omitted mention in his instructions of more complicated navigational technologies, such as those for measuring and calculating the magnetic variation of the compass. Knowing the variation should have been particularly relevant for the Arctic explorers, given the high latitudes in which the voyage was to take place and the importance of accurate compass bearings in compiling new charts and rutters.[32] Yet Cabot, if indeed he himself knew how to take such measurements, may have felt that they were simply too much for his inexperienced pilots to handle. None of them, after all, had any significant experience in using mathematical technologies of navigation or even possessed the physical means to do so before 1553. Although sixteenth-century pilots customarily supplied all of their own navigational equipment, few if any of Cabot's English mariners yet owned such unfamiliar devices, and so the instruments for the maiden voyage were provided instead "at the charge of the company."[33]

Besides Cabot, a number of native English mathematicians were active, directly or indirectly, in advising the explorers. Although there is no solid evidence of Robert Recorde's formal involvement with the Muscovy Company before 1553, in 1551 he published the second in his series of popular vernacular textbooks in mathematics, *The Pathway to Knowledge*, a dialogue teaching basic geometry. In his preface to the book, in which he discussed "the commodities of Geometry, and the necessity thereof," Recorde argued that geometry was essential both for shipbuilding and for creating navigational instruments such as the compass and chart.[34] In later years Recorde's connection with the Muscovy Company became more overt; he even dedicated his 1557 textbook on advanced arithmetic, called *The Whetstone of Witte*, to "the right worshipful, the governors, consuls, and the rest of the company of venturers into Muscovia." In the dedication he promised the governors of the company that he would soon write an entire treatise on the art of navigation, "for your pleasure, to your comfort, and for your commodity," focusing especially on "the northly navigation," though unfortunately no such work survives.[35]

The mathematician, geographer, and astrologer John Dee also helped to prepare the explorers for their voyage. During the late 1540s, Dee had traveled across western Europe, seeking and winning the friendship of such geographic and mathematical luminaries as Gemma Frisius, Pedro Nuñez, and Gerardus Mercator. Upon his return to England in 1550, he not only brought with him the latest Continental maps, globes, and instruments but also began writing treatises on the use of the globe, on meteorology, and on astronomy for his patrons in the powerful Dudley family, who were active supporters of, and investors in, the Muscovy Company.[36] Dee not only placed his growing library, his charts, and his considerable personal knowledge at the disposal of the company; he also began a friendly collaboration with the pilot major of the fleet, Richard Chancellor, who in 1551 had gained some practical experience at sea on Bodenham's Levant voyage.

Little is known of Chancellor's education or background before 1551, but he was born in Bristol and probably trained as a mariner, though at some point he also came under the patronage of Sir Henry Sidney, who educated him and brought him to the Muscovy Company's attention. In any case, Chancellor was certainly no common mariner but an able mathematician and a fine instrument maker, who excelled at modifying extant instruments in order to suit his own needs. His mathematical talents were such that Dee, a mathematician respected throughout Europe, treated him as a peer. They worked jointly in 1553, on the eve of the voyage, to produce a new ephemerides of astronomical observations, probably intended for navigational purposes, using instruments modified by Chancellor. Many years after the latter's death, Dee still referred to him as "the incomparable M. Richard Chancellor."[37] Although Sir Hugh Willoughby, as a career military officer, was ultimately named captain general of the mission, "both by reason of his goodly personage (for he was of a tall stature) as also for his singular skill in the services of war," it was Chancellor in the office of pilot major, "a man of great estimation for many good parts of wit in him . . . in whom alone great hope for the performance of this business rested."[38]

In the event, the "great hope" of the company was not misplaced. The tragic end of Willoughby and his company has already been recounted. Chancellor's ship, however, sailed as far as the White Sea, where he made landfall and was received by the Russians as an English envoy and conveyed overland to Moscow. There, even in Willoughby's absence, Chancellor so impressed Czar Ivan IV with his conduct that the latter awarded the Muscovy Company exclusive trading rights at Russian ports. Having returned to England in triumph in 1554, Chancellor was sent back to Russia by the company (now formally chartered and incorporated by Queen Mary) in the following year and was to carry the first Russian ambassador to England upon his return in 1556. However, his returning fleet of four ships (two of which were the same ill-fated ships lost in Willoughby's original voyage, discovered in the interim by the Russians) was scattered by rough weather near Scotland. Three ships were lost, including Chancellor's, and the pilot major himself drowned off the Scottish coast in attempting to convey the ambassador and his entourage to shore in the ship's boat.[39]

The methods Chancellor used in navigating from London to the White Sea are not known. Given his recognized mathematical talents, he may well have brought to bear some or all of the new technologies that Cabot and Dee could teach him. But Chancellor was also acknowledged by his contemporaries to be a special case. Given Hugh Willoughby's nonmaritime background and the short time that he was actually in Chancellor's company at sea, the ships of the fleet other than Chancellor's were far more likely to have relied upon traditional dead reckoning, as best they could in unfamiliar waters. Willoughby himself recorded mostly traditional piloting data in his journal—the compass headings followed and the estimated distances sailed—though the ships' course is all but impossible to determine given the journal's confused state. The journal does contain some scattered latitude measurements, indicating that Willoughby or some of his officers attempted to measure the altitude of the sun or the pole star; but the last several entries describe them sailing all over the Barents Sea, unable to find even the known safe harbors at Vardø or on the White Sea.[40] It is perhaps unsurprising that Willoughby froze to death in Lapland while the more skillful and experienced Chancellor survived the winter and returned home safely the following spring. And although he, too, perished during his next voyage, his initial success was a pregnant one; Chancellor's promising young ship master in 1553, Stephen Borough, went on to become one of the most influential figures in English navigation over the next two decades.

## Stephen Borough, "chief pilot . . . of England"

Stephen Borough's navigational duties as the master of Richard Chancellor's ship are unclear. As stated above, the master was typically the officer responsible for setting and maintaining the ship's course, but Chancellor's presence on board as the fleet's pilot major would seem to make Borough's role redundant. The relationship between the two officers was probably pedagogical, with Borough taking advantage of a valuable opportunity to observe and learn from an acknowledged master under actual maritime conditions. Nothing is known of Borough's background or training before his voyage with Chancellor in 1553, though he soon proved to be a talented navigator, quickly winning the trust of his employers in London. In 1556, just as Chancellor was beginning his fatal return trip to England with the Russian ambassador, the Muscovy Company gave Borough command of his own voyage: a mission to explore as far as the River Ob' in northern Russia. Borough himself wrote a vivid account of the voyage, later printed by Richard Hakluyt in his *Principall Navigations*, thus preserving one of the earliest and most detailed records of the use of mathematical navigation technologies by an English navigator.

Although Borough was to be responsible for escorting an English merchant ship as far as Vardø, his primary mission was not trade but exploration; rather than taking a whole fleet of ships laden with merchandise, as Willoughby had, Borough commanded only a single small vessel called a pinnace. In the sixteenth century, pinnaces most often accompanied larger merchant vessels in a support role, scouting ahead for shoals and sufficiently deep channels through them. Designed to be maneuverable in tight spaces, the pinnace proved to be well suited for the exploration of unknown Arctic seas. Before their departure from the port of Gravesend on 1 May 1556, Borough and his crew were feted by Sebastian Cabot himself, who had just been appointed governor of the Muscovy Company for life. Borough wrote that Cabot, by then an old man, even "entered into the dance himself, amongst the rest of the young and lusty company; which being ended, he and his friends departed most gently, commending us to the governance of Almighty God." Cabot almost certainly did not survive to see Borough's safe return to England, but it must have gratified him to see the endeavors of his last years come to fruition in one of his young protégés.[41]

As he kept track of his navigational data throughout the voyage, Borough always endeavored to be as thorough and precise as possible, and perhaps even a bit more. His weekly latitude measurements, for example, were always recorded to the nearest minute of arc—a remarkable, and indeed dubious, degree of precision, vastly greater than that of the contemporary solar declination tables used in making the latitude calculation itself.[42] Five days out from Gravesend, Borough's journal records a milestone in the history of English navigation: the first logged observation of magnetic variation of the compass. Willoughby's journal, in contrast, contains no mention at all of magnetic variation, and though Chancellor might have made such measurements, no references to them survive. During the sixteen months he was away from England, Borough made six recorded measurements of magnetic variation, some taken on shore and some while under way. Making these observations was not difficult in principle, but it could be challenging aboard a moving ship and almost certainly required two men: one to measure the height of the sun to determine the exact time of local noon, and one to compare the direction of the sun's shadow at noon (due north) with that of the compass needle, all on the deck of a pitching and rolling ship. It is no wonder that Borough settled for the nearest half-degree of arc in recording magnetic variation.

Despite Borough's proficiency in both traditional and mathematical navigation, not all of the careful piloting exhibited during the voyage was his own, though for much of the trip he did the best thing a smart pilot could do: he found someone else who knew the seas and coasts better than he did, and he followed him. Early in his mission, Borough was fortunate enough to make contact with a small fleet of Russian fishing vessels that were heading in the direction he wanted to go. After he befriended and entertained some of the Russians aboard his ship, they agreed to hang back with him to help guide his pinnace, though their fishing boats were smaller, faster, and more maneuverable. The Russians accompanied Borough from 11 June until 6 August 1556, when they passed over shoals too shallow for his own ship to cross. On 22 June, one particularly helpful Russian pilot named Gabriel abandoned the company of his fleet to guide Borough and his crew safely over some especially treacherous shoals; for his trouble, Borough gave him "two small ivory combs, and a steel [looking] glass, with two or three trifles more, for which he was not ungrateful." On 24 June, Gabriel and the other Russians managed to keep Borough's ship from beaching on a sandbar in rough weather by loaning the

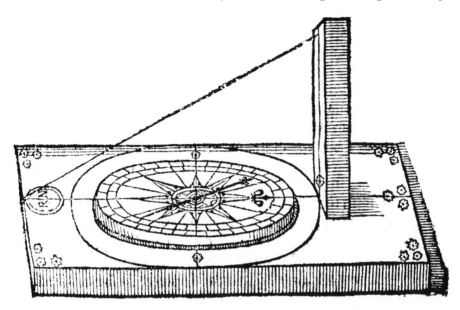

*Figure 3.7.* A compass of variation, used to measure the magnetic variation. The navigator oriented the compass along a precise north-south axis, according to the shadow cast by the gnomon at exactly local noon. Then, in comparing the shadow's heading with the direction indicated by the compass fly, he could observe the angular discrepancy between true and magnetic north. William Borough, *A Discovrs of the Variation of the Cumpas, or Magneticall Needle*, 2d ed. (London, 1585), fol. B.ii r. Photo courtesy of the Beinecke Rare Book and Manuscript Library.

English a couple of anchors, and they even salvaged a hawser the English lost in the incident.[43]

In addition to shoals and sandbars, the English mariners faced some hazards peculiar to Arctic navigation: on 21 July, at the height of summer, they found themselves trapped in an ice floe. Ice could be an especially dangerous threat for sailors, because it constantly shifted and could rapidly encircle a ship, as Borough described: "Within a little more than half an hour after we first saw this ice, we were enclosed within it before we were aware of it, which was a fearful sight to see: for, for the space of six hours, it was as much as we could do to keep our ship aloof from one heap of ice, and bear roomer from another, with as much wind as we might bear a course." They also had a close encounter with "a monstrous whale" on 25 July, "so near to our side that we might have thrust a sword or any other weapon in him, which we durst not do for fear he should have overthrown our ship." After some vigorous yelling on the part of the entire crew, "God be thanked, we were quietly delivered of him."[44]

Despite making excellent progress in their journey, Borough and his crew failed to reach the River Ob' that summer. By the middle of August the Arctic nights were getting steadily longer, the storms more frequent and severe, and the ice floes more ubiquitous and threatening. Borough prudently decided to winter in a haven he knew rather than freezing in the ice as Willoughby had. After spending the winter at the Russian port of Kholmogory, instead of setting out for the Ob' again as he had intended, in the spring of 1557 Borough was instructed by agents of the Muscovy Company to sail to Vardø and attempt to discover what had become of the fleet of ships in Richard Chancellor's last voyage. The only news heard in London to that point was of Chancellor's wreck, of which they had learned from the few survivors. The fate of the remaining three ships was still a mystery, as even the single surviving ship had not yet returned to England, having wintered instead at Trondheim.

Borough did eventually learn the fates of all the company's missing ships, but his frequent and careful navigational observations ultimately did the shareholders a far greater service. As he crisscrossed throughout the Russian Arctic in search of Chancellor's lost fleet, Borough kept such a thorough and meticulous log that at the end of his journey he was able to compile a complete rutter for sailing the Arctic seas northeast of England. His new rutter listed the headings and distances between more than a dozen ports, havens, capes, and islands between Vardø and the White Sea, where scores of English ships would soon sail in a regular and profitable trade with the Russians. Throughout, Borough provided useful information regarding tides, water depth, and the locations of the many underwater hazards he encountered, as well as several latitude observations and a few magnetic variation measurements. And while his seaman's tales of ice floes, whales, and other disasters narrowly averted certainly added color and drama to his narrative, they also served as a warning and example for the English mariners who would follow him, illustrating how to sail in the frigid Arctic seas with confidence and success. He even included a very brief but practical Russian dictionary, containing common words and phrases of nautical and mercantile interest, such as *sun, moon, stars, rain, ice, stone, wind, friend,* "What call you this," "Come hither," "Good morrow," "I thank you," and numbers as high as one hundred—all of which he had learned, no doubt, from his benevolent Russian guides.[45]

In short, by providing such a wealth of information from his own experience, Stephen Borough helped to transform the trip to the White Sea from a voyage of exploration into a normalized trade route. As a result, he ceased to be simply a navigational practitioner and became an expert mediator. Rather than maintaining craft secrecy and exploiting his exclusive possession of such valuable experience, as a traditional pilot might have done, Borough opted to share his knowledge openly with the entire Muscovy Company. This act brought Borough great public renown as an explorer and earned him the gratitude of his patrons and employers, even as it rendered his remarkable achievement attainable for other English mariners. Although the initial voyage had required the skills of a gifted, experienced, and mathematically educated navigator, Borough's creation of a new rutter placed the same journey within the reach of any well-trained English pilot—further illustrating the difference between exploratory navigation and the more traditional styles.

The following year, Borough traded upon his greatly enhanced reputation, taking an opportunity to enhance it still further; he temporarily gave up voyaging to the northeast to travel south for a change. Queen Mary's marriage to King Philip II of Spain in 1554 had cemented Anglo-Spanish relations to the point where Borough was welcomed at the Casa de Contratación in Seville as an esteemed guest. He was reportedly received there with "great honor" and was even presented with "a pair of perfumed gloves worth five or six ducats." Bringing Borough to the Casa was probably no act of charity on the part of the Spanish; they were well aware of his Arctic voyages and "[had] intelligence that he was master in that discovery." The cosmographers at the Casa no doubt hoped that Spain would benefit from Borough's rare navigational experience in that area. How much they gleaned from him about his Arctic travels is unknown—Spain did not pursue any Arctic explorations of their own. Borough, however, took away a great deal from his time in Seville. He was present for the "making and admitting of masters and pilots" at the Casa and, like Sebastian Cabot before him, had access to the very latest Spanish navigational techniques, instruments, charts, and manuals.[46] The exact nature of his participation in the Casa's affairs is unknown, but he certainly observed them approvingly and sought to bring Spanish navigational technologies and training methods back with him to England.

Once again, rather than exploiting his unique experiences in Spain solely for his own benefit, upon his return to England in 1560, Borough chose to share what he had learned with his colleagues. His goal was to ensure that English pilots caught up as fully and rapidly as possible with what he believed to be a superior Spanish system. Having brought back with him a copy of Martín Cortés's navigational training manual *Breve compendio de la sphera y la arte de navegar*, published in Spain in 1551, Borough donated the book to the Muscovy Company and recommended that they have it translated for the use of their pilots. The governors of the company duly secured the services of Richard Eden, the Cambridge-educated translator, amateur geographer, and William Cecil's private secretary; in 1561 Eden published the first English-language navigation manual, *The Arte of Nauigation*.[47]

In his dedication of the translation to the Muscovy Company's governors, Eden drew a sharp distinction between the traditional art of piloting and the new mathematical navigation technologies to which Borough sought to expose English pilots. He disdained the locally bound knowledge of "certain fishermen that go a trawling for fish in catches or mongers," never even venturing beyond the Thames estuary, and lauded instead "such excellent pilots as are able without any rutter or card of navigation, not only to attempt long and far voyages, but also to discover unknown lands and islands." Stephen Borough, Eden argued, should be considered a hero among English mariners and richly rewarded by his mercantile employers: "For certainly when I consider how indigent and destitute this realm is of excellent and expert pilots, I can do no less of conscience than . . . to exhort you and put you in remembrance . . . so to regard him and esteem him and his faithful, true, and painful service toward you, that he may thereby be further encouraged, and not discouraged."[48]

Cortés's manual, as Eden rendered it, was illustrative and helpful in some ways, yet confusing in others. It emphasized navigational theory as well as (and sometimes at the expense of) practice, in the belief that the reader should understand the basic cosmological and mathematical principles at the root of this newly enhanced art. Cortés had included (and Eden preserved) a number of visual and textual aids throughout, in the form of simple diagrams, astronomical tables, and even movable volvelles, which allowed the reader to use the book itself as a calculating device for some tasks; the book therefore served as a mathematical instrument as well as an instructional text. Its greatest overall strength was its clear instructions for constructing one's own navigational instruments, including the mariner's astrolabe, cross-staff, sundial, nocturnal, compass, and even plane charts.[49] These, taken together with the various instruments and tables included within the book itself, and in conjunction with the sort of institutionalized instruction that Cortés could assume Spanish pilots would receive at the Casa de Contratación on how to use all of the information and equipment he discussed, probably formed an excellent foundation for an aspiring Spanish pilot in 1551.

This did not mean, however, that the book translated well into the English context of the 1560s. At that time, no English pilot had access to anything like the established training and examination system found at the Spanish Casa. Stephen Borough's time there was virtually unique, as the privilege was not extended to other English pilots after Queen Mary's death and the Protestant Queen Elizabeth's succession to the throne in 1558. Whatever navigating an English seaman might know how to do he had learned by observing his officers at sea and then doing it himself as a pilot's apprentice or master's mate. With Cortés's book in hand,

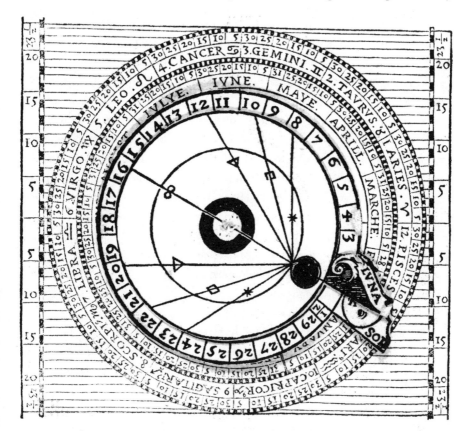

*Figure 3.8.* A multipurpose textual instrument in Cortés's popular navigation manual, *The Arte of Nauigation*. This instrument could be used to determine the declination of the sun, the phase of the moon, and the moon's location in the zodiac, among other things. It was one of the most elaborate in the manual, with two nested, rotating volvelles. Martín Cortés, *The Arte of Nauigation*, trans. Richard Eden (London, 1561), fol. xxxii v. Photo used by permission of the British Library, shelfmark G.7310.

an aspiring (and confidently literate) English pilot would have been able to name, and perhaps even to build, several navigational instruments—and yet still be utterly unable to use them successfully. Cortés's manual, taken by itself, was insufficient for training mathematical navigators. If the English were ever to master the new mathematical technologies employed by the Spanish, Borough realized, they would need further instructional support and encouragement—and he soon approached England's new queen with this goal in mind.

Queen Elizabeth, her new Privy Council, and the English Parliament of the early 1560s all actively supported English maritime development and practical reform. As Henry VIII had promoted English maritime self-sufficiency, so did Elizabeth and her government: by 1565 the sale of English ships to foreigners had been banned; small craft were forbidden from pursuing foreign trade, necessitating the design and construction of larger merchant vessels; the profitable French wine trade was limited to English ships with English masters, so that native officers might acquire some valuable overseas experience; ship owners and masters were formally empowered to take on apprentices, whereas this had previously been only an informal arrangement; statutory fish-eating days were instituted, and English-caught fish could be landed duty-free, in order to encourage the fishing trade; artificial buoys and beacons were erected to assist coastal pilots, and it was forbidden to tear down or remove prominent trees or church steeples that were recognized as coastal landmarks; and the coasting trade itself was also confined to English ships and crews, so that foreigners would not have such easy opportunities to learn the English coastline for potentially hostile purposes.[50] In the spirit of these various resolutions, Stephen Borough petitioned the Crown in 1562 to create and fund the office of pilot major of England.

The principal duty of the Spanish pilot major at the Casa de Contratación was to train and license every Spanish pilot conducting voyages to the New World. No Spaniard was permitted to call himself a pilot or ship's master unless he held a certificate from the pilot major granting him that privilege, which he could only earn by passing an examination. If an aspiring Spanish pilot was examined and found wanting, the Casa could send him back to sea in an apprenticeship capacity until he was ready to be reexamined. Borough saw several advantages in such a system. It provided a means to "make perfect mariners," rather than allowing "accustomed ignorance" to be perpetuated, which in turn brought "great benefit, honor, and fame" to Spain and prevented "losses of ships or shipwreck through ignorance of mariner craft."[51] The Casa's centralized training and examination system, according to Borough, kept incompetent pilots from slipping through the cracks and endangering ships and crews; although maintaining such an institution was expensive, it was supposed to prevent shipping losses and save lives.

In England, by contrast, Borough painted a grim picture of ignorance and disorganization. The Spanish, he wrote, maintained a structured hierarchy of navigational experience and expertise, through which every ambitious mariner had to progress, including "the pilot, the master, the mariner, the grummet, the page, and the boy," with all advancement regulated by the Casa. The English system was much less regimented: "we have not ... in our English ships but man and boy." Career advancement was based solely on one's length of service at sea, and not on any formal navigational knowledge or skill. "[A]s soon as a young man cometh to any reasonable stature," Borough lamented, "he will look for his age and not for his knowledge to have the name of a man and also of a mariner although he understand little in the art."⁵²

As for the prevention of shipwrecks and the honor of their country, Borough argued that the Spanish pilots were so successful and well respected precisely "because of their daily exercise in the principles of the art" of navigation, which for him included a mastery of the new mathematical technologies. Successful Spanish pilots were required "to know the latitude [by] the sun or stars, the variation of the compass, [and] diverse other sundry rules and ways whereby they know and reckon their ships' ways exactly." In contrast, these "exact rules and reckonings" were understood by "the fewest sort of mariners in this noble realm of England. . . . the greatest number of our English mariners contenteth themselves with the old ancient rules, as they term it, which is erroneous enough." Some English pilots, Borough suggested, were both aware and ashamed of their deficiencies, but they had no means of correcting them. In order that "they would not seem to be unskillful or ignorant in the principal rules of navigation," these men took mathematical instruments to sea with them simply for show, "although they be utterly ignorant in any use of the same." Still others, overconfident in their small knowledge, became a danger to themselves and others: "when they understand a little, they think they know all . . . And hath been the chiefest occasion of the casting away of many men's lives as well as the losses of ships and goods." As a result, English ship owners were reduced to begging abroad, looking to France and Spain "for men skillful in that art."⁵³

Borough considered the traditional art of piloting by "the old ancient rules" no longer sufficient to meet the new challenges that English mariners faced. Antwerp and Seville were slipping from their former prominence as the main English ports of call, and sailing across uncharted waters in search of new markets, Borough believed, would require a mastery of mathematical navigation technologies not previously used by most English pilots. In order to introduce these technologies into a community of conservative practitioners who knew little or nothing about them, Borough realized that a centralized system of instruction would be most efficient. He therefore asked the queen and her Privy Council for permission and funding to erect a rigorous, centralized, institutional training system in England on the model of the Spanish Casa de Contratación. Through such an organization, Borough hoped to ensure that his own ideas about the required skills and qualifications for a practicing pilot would become standard—indeed, the only permissible ideas—throughout the English maritime community, for anyone hoping to obtain an English pilot's certificate. Though Borough did not go so far as to suggest himself for the job, as the only living Englishman with any firsthand knowledge of the Spanish office he hoped to emulate, he would have been the only obvious candidate for the new post of pilot major, and he was surely aware of this. With his petition, therefore, Stephen Borough sought nothing less than a royal mandate to define and propagate navigational expertise in England by himself, to fashion all future English navigators in his own image.

He very nearly got what he wanted. In January 1563 the Privy Council drafted a document appointing Stephen Borough as "the chief pilot of this our realm of England." Borough was to be given not only control of "the examination and appointing of all such mariners as shall from this time forward take the charge of pilot or master upon him, in any ships within this our realm" capable of carrying more than forty tons, but also the right to approve the appointments of virtually all other ships' officers, including boatswains, quartermasters, and masters' mates. Furthermore, the grant formally delineated a hierarchy of inferior stations aboard ship, adding the ranks of page and grummet between those of mariner and boy. In order to become a ship's master, then, an aspirant would have to work his way up the full ladder of ranks and then, upon examination, receive a seal of approval from the chief pilot; for piloting without a seal, he would be liable to a fine roughly equivalent to a month's wages for a mariner. In addition to examining and licensing all English pilots, Borough was to be entitled to set the training curriculum, to "give rules and instructions touching the points of navigation and at all other times to be ready to inform them that seek knowledge at his hands."[54]

Borough's appointment, if instituted, would have made the training, selection, and promotion of the entire officer corps of every ship in England a royally sanctioned and monitored activity, carried out centrally in London by a pensioner of the Crown. Such a move was unprecedented in English maritime history and would have been a major blow to the traditional powers of local maritime authorities, at ports scattered throughout the realm. Borough's authority as chief pilot of England certainly would not have gone unchallenged at the local level, though as it happened no challenges arose because the Privy Council never went forward with his appointment. The document in question was only a draft; crucial pieces of information (such as the chief pilot's salary) were left blank, no finished, formal version of the appointment survives today, and no subsequent mention of Borough as the "chief pilot of England" has been found.[55]

Yet despite the overt failure of his petition, Borough managed to wield considerable power over English navigational training for decades to come, by acquiring a trio of more modest, already extant offices and exercising their powers collectively. In 1565 he became one of the four masters of the royal ships in the Medway; in addition to overseeing the maintenance of the ships in the naval fleet, the masters of the royal ships also examined and selected pilots and masters for them. Though the post was not as exalted or comprehensive as that proposed by Borough in his petition, it was vital for the well-being of the Royal Navy, and it put Borough near the center of naval administration. The royal ship masters also worked closely with Trinity House of Deptford Strand, a fraternal organization founded under Henry VIII for the purpose of ensuring a ready supply of trained pilots capable of guiding ships safely in and out of difficult havens along the Thames estuary. In addition to his naval post, Borough became a member of Trinity House; within a decade he was elected its master and quickly set about expanding the organization's mandate. In a parliamentary act of 1565, "concerning sea-marks and mariners," Trinity House was granted the authority to issue certificates of competence to any mariners sailing from or within the Thames estuary. This was very probably at Borough's urging, either directly or indirectly through his petition of 1562, and ultimately gave him a second administrative post from which to train and examine English pilots sailing in and out of England's busiest ports. Finally, Borough remained the chief pilot of the Muscovy Company, a post he had acquired upon Chancellor's death in 1556; this gave him an important voice in selecting the pilots who would guide the company's ships in all of their trading and exploratory endeavors. The company held a royal monopoly on all trade and exploration to the north, northwest, and northeast, so Borough had a third post from which to oversee personally the training and selection of ambitious English pilots, especially those operating from the Thames estuary.[56]

Although this trio of offices did not give Borough all the centralized, blanket powers he had originally petitioned for to train and license pilots throughout the realm, taken together they provided him with a reasonable substitute. By 1565 Borough had acquired near total control over the instruction and hiring of pilots and ships' masters employed by the Royal Navy, the Muscovy Company, and the other London merchants—and all without creating or funding any new offices or institutions or provoking challenges from local maritime authorities. As for the knowledge and skills Borough expected his pilots to possess, unfortunately there are no records of the sort of training he may have offered. Given his active encouragement of Richard Eden's translation of Martín Cortés's navigational manual, however, we may assume that his training regimen would have required a basic understanding of the technologies covered in that text. Thus, within just a few years after his return from Spain, Stephen Borough had managed to set himself up as the foremost navigational expert in England, with the authority both to require and to provide practical training in mathematical navigation for any pilot or ship's master who sailed from England's busiest ports of call, London and the other harbors of the Thames estuary. Over the succeeding decades, Borough's mathematical training program permanently altered English navigational practice, at least with respect to voyages of exploration.

## Training Arctic Explorers: Frobisher, Pett, and Jackman

Until his death in 1584, Borough continued to hold the three supervisory offices he had acquired and worked to make his conception of navigation the English standard. For the rest of the sixteenth century, the main elements of his mathematical navigational program were discussed and developed in numerous English navigational manuals (several of which are examined in chapter 4) and can also be traced through surviving documents concerning the preparations for voyages of Arctic exploration. Borough's navigational priorities were much in evidence, for example, during the hurried preparations for the first of Martin Frobisher's three northwestern voyages in 1576, and Borough himself was one of the navigational and geographic experts called upon to assist in them.

Frobisher was born during the mid-1530s in Yorkshire. His father died during Martin's infancy, and "for lack of good schools thereabout," he was sent at an early age to London to be raised by a kinsman, Sir John York. York determined that Frobisher was well suited for a nautical career and sent him on his first mercantile voyage to Guinea in 1554—just three years after Bodenham's path-breaking voyage to the Levant and one year after Willoughby's attempt to the northeast, putting Frobisher near the forefront of the English drive to reach new overseas markets.[57] He acquired a great deal of maritime experience over the next decade, sailing mostly on the increasingly common trading voyages to northern Africa and the eastern Mediterranean. In 1566 he was arrested and questioned under suspicion of having outfitted a vessel with the intent to commit piracy—a common enough activity among English mariners during Elizabeth's reign, as often as not with the queen's blessing, or at least her forbearance. Indeed, by 1571 Frobisher was in sufficient favor at court that Lord Treasurer William Burghley had a ship outfitted for him at the Crown's expense, for unknown purposes.[58]

By 1574 Frobisher had become interested in searching for a northwestern trade route to Asia, in opposition to the contemporary English fixation on the northeast. When he applied to the Privy Council for permission to conduct a search, the councillors informed him that the Muscovy Company held a monopoly on all northern exploration and he would have to secure their approval first. Frobisher then met with several representatives of the company, including their chief pilot,

Stephen Borough; but they refused to back his venture, believing that the northeastern route held more promise and, in any case, was already half explored. Frobisher continued to agitate, however, and gained the sympathy of the Privy Council. Lord Burghley himself finally ordered the governors of the Muscovy Company either to back Frobisher or else to license him to make the attempt using whatever funds he could raise on his own. The company took the latter option, minimizing their own risks and expenses.[59]

Frobisher soon formed an independent partnership with Michael Lok, a prominent London merchant who called himself an agent of the Muscovy Company, "having the charge of all their business to understand the ground of this case."[60] Lok had been sent by his father to the Continent at the tender age of thirteen, so that he might gain some practical experience in the family's mercantile business. He wrote that he spent the next fifteen years traveling through Scotland, Ireland, France, Flanders, Germany, Spain, Italy, and Greece and also claimed to have commanded a cargo ship of one thousand tons for three years during the last days of the English Levant trade. This would make him a most experienced merchant and ship's master by contemporary English standards. During his extensive travels, Lok also became an amateur geographer, studying "histories in many tongues" and spending more than five hundred pounds on "books, maps, cards, and instruments . . . for mine own private study [and] satisfaction in the knowledge of this matter." His own collected geographic notes, he wrote, would have filled "a ream of paper."[61] Early in his career, Lok had been struck by the immense wealth Spain and Portugal were amassing in their trade with the East Indies; his intent in supporting Frobisher in the 1570s was to bring a share of those profits to England.[62]

In addition to his financial support, Lok worked hard to prepare Frobisher for the navigational challenges he could expect to face in his first voyage of exploration, claiming that he had personally "instructed him to my skill, showing him all my books, maps, sea-cards, instruments, and notes made."[63] Nor did Lok have to rely solely upon his own resources; on 20 May 1576, shortly before Frobisher's planned departure, John Dee heard about the preparations for Frobisher's voyage through "the common [report]." Dee approached Lok himself and offered to help instruct Frobisher "with such instructions and advice as by his learning he could give therein."[64] Appreciative of the offer, Lok hastily arranged a meeting at which Frobisher and another ship's master on the voyage, named Christopher Hall, were given a sort of "masters' class" in mathematical navigation by John Dee, Stephen Borough, and himself.

Also assisting in Frobisher's preparations was William Borough, the younger brother of Stephen and the same master mariner who would soon become involved with the survey and rebuilding of Dover Harbor. William Borough had first sailed with Richard Chancellor as ship's boy in 1553, at the age of sixteen, and again under his brother's command in 1556. By 1576 he was a highly experienced ship's master and navigator in his own right, having made both mercantile and exploratory voyages to Russia nearly every year for the Muscovy Company; his geographic knowledge of the northeastern Arctic seas was unsurpassed.[65] Although not confident enough in Frobisher's chances for success to invest his own money in the voyage, he nevertheless assisted in the preparations by finding a ship's master and several mariners to fill out Frobisher's crew, making several suggestions about equipment to include and consulting with Frobisher about his intended course.[66]

Michael Lok, John Dee, Stephen and William Borough: such a group can only be described as a remarkable panel of navigational instructors, quite unprecedented in English maritime history. Dee in particular was one of the most respected mathematical and geographic authorities in Europe, and between them the Borough brothers had decades of experience in applying mathematical navigation technologies in Arctic waters. But then, Martin Frobisher himself was certainly no green hand aboard ship. By 1576 he had spent most of the previous twenty-one years at sea, often as the master of his vessel, and his considerable maritime experience had attracted the attention and confidence of Lord Burghley and other members of the Privy Council. Why would such an experienced seaman and ship's master have needed such an impressive battery of navigational instruction?

The answer was that Frobisher, for all his two decades of nautical experience, probably lacked the one thing common to almost everyone else involved with the preparations for the voyage: a mastery of mathematical navigation technologies.[67] The majority of Frobisher's previous voyages had been along established trade routes in comparatively well-known bodies of water, such as the Mediterranean Sea and along the West African coast. Although some of these routes may have been new to the English, that hardly mattered when English ship's masters still commonly hired (or kidnapped) Iberian pilots to guide them through unfamiliar waters. Frobisher might very well have performed all the maritime duties expected of him up to 1576 and still never have learned the mathematical skills that someone like Stephen Borough considered to be of paramount importance, especially for an explorer. The navigational challenge of sailing into unknown and uncharted Arctic seas was certainly unlike anything Frobisher had ever faced. And he owned none of the equipment necessary for mathematical navigation—Lok and the other investors in the voyage spent just over fifty pounds to provide him with new navigational instruments and manuals. This was no trivial sum; in comparison, the investors paid £120 for the ship *Michael* and all her furniture and a little more than £152 to have the ship *Gabriel* and an accompanying pinnace newly built for the voyage. The amount spent on new navigational equipment therefore represented well over one-third of the total value of either thirty-ton ship when newly built and more than 3 percent of the total cost of outfitting, manning, and provisioning Frobisher's first voyage.

What exactly did the investors get for their fifty pounds? Michael Lok purchased the instruments himself, and his choices included a blank metal globe for making navigational calculations; a brass armillary sphere for astronomical calculations; a meridian compass for finding magnetic variation; a plane table with sights and a standing level for surveying and mapping the coastline; a sundial for use in any latitude, called a universal ring dial; several more brass astronomical instruments; a wooden cross-staff; an astrolabe, purchased from William Borough; cases for most of the instruments; and a large assortment of ships' compasses (twenty) and hourglasses (eighteen). Lok also purchased a wide variety of maps and charts, including "a very great card of navigation," which cost five pounds by itself; the 1569 printed world map of Gerardus Mercator; another large printed map by Ortelius; three more small printed maps; and six manuscript charts (among which were England's earliest experiments with the circumpolar projection, supplied by John Dee) for recording discoveries along the way. Finally, the fleet was equipped with a small library, including Robert Recorde's *The Castle of Knowledge* (1556); William Cuningham's *The Cosmographical Glass* (1559); *The new found vvorlde, or Antarctike . . .* by André Thevet (1568), as well as his *La cosmographie universelle*, in French (1575); Pedro de Medina's *Regimie[n]to de nauegacio[n]*, in Spanish (1563); a medieval travel narrative attributed to John Mandeville (1568); and an English Bible, of "great volume."[68] The items on Lok's shopping list represented the cutting edge of geographic, cosmographic, and cartographic knowledge with respect to the art of navigation in 1576, giving credence to Lok's claims to expertise in those subjects. Of this list, the only items Frobisher was likely to have encountered in his previous travels were the compass, the hourglass, the plane chart, the cross-staff, and the astrolabe (and perhaps the Bible)—only the most basic mathematical navigation technologies, carried aboard every Iberian ship by the early sixteenth century.

More than twenty years' experience as an English merchant seaman were not enough to qualify Frobisher as a navigator fit for Arctic exploration, according to the standards set by Lok, Dee, and the Borough brothers. The very fact that the voyage's investors had to supply such expensive navigational equipment out of their own operating capital, rather than relying on Frobisher and the other ships' masters to supply their own equipment in traditional fashion, indicates that they expected more of Frobisher than he was accustomed to doing. It recalls the Willoughby voyage of 1553, for which the Muscovy Company had also supplied all navigational instruments, since no English pilot could yet be expected to own them himself. And if Frobisher had to be supplied with the requisite tools for mathematical navigation, he would most likely also have needed proper instruction in how to use them. In considering the challenges before him, Frobisher himself may even have been conscious of his navigational shortcomings: John Dee wrote in 1577 that he initially got involved with the venture in part because Frobisher had personally requested his help and instruction.[69]

Whether or not Frobisher was able to apply whatever he had managed to learn in his crash course in mathematics and geography to find his way at sea is debatable; a few days was not a very long time to practice when trying to master tricky and unfamiliar navigational technologies. On the one hand, George Best, one of Frobisher's officers and the author of a firsthand account of the voyage, wrote that "our worthy Captain *Martin Frobisher* . . . hath diligently observed the variation of the [compass] needle."[70] Frobisher almost certainly never took such measurements in the Mediterranean or eastern Atlantic, where the trouble caused by the phenomenon is minimal, so he probably learned the technique from the Borough brothers specifically for his Arctic explorations.[71] On the other hand, during his third voyage in 1578, Frobisher got dangerously lost in searching for what should have been by then a familiar harbor in North America. Stubbornly insisting that he was in the correct place and overruling the objections of the other ships' masters, he cavalierly ordered most of his fleet into an unexplored bay where only blind luck prevented them from running aground.[72] This potentially fatal blunder, shocking in such an experienced officer, could probably have been avoided by making a single accurate latitude calculation, and Frobisher's failure to do so reflects poorly upon his competence as a mathematical navigator.

In any case, Frobisher and Hall did express gratitude for their brief instruction. Shortly after their voyage began, from a bay in Scotland where they had stopped to repair a leak, the two veteran mariners wrote to John Dee in London, offering him "as many thanks as we can wish" for his "friendly instructions." Although Frobisher indicated that they were trying to apply what they had been so hastily taught, he also admitted that they had not been the ready pupils Dee might have wished for and apologized that he and Hall were but "poor disciples, not able to be scholars but in good will for want of learning." Quoting from the letter in a later work, Dee went on to state that both mariners "became very sorry of their so late acquaintance and conference . . . and greatly misliked their want of time."[73]

Frobisher's mathematical preparations were not unique; subsequent Arctic explorers received similar barrages of navigational instruction before their voyages, despite having considerable maritime experience. In 1580, two years after Frobisher's last attempt to find a northwestern passage had come to naught, Charles Jackman and Arthur Pett prepared to make yet another exploratory journey to the northeast for the Muscovy Company. Like Frobisher, both men were experienced mariners, and both had been on voyages to the Arctic before: Pett had sailed with Richard Chancellor and the Borough brothers in 1553 and had become a regular ship's master for the Muscovy Company, whereas Jackman had been Christopher Hall's mate in 1576 and then a ship's master in his own right on the later Frobisher voyages.[74] Each man was also presumably familiar with the basics of mathematical navigation; both sailed from the Thames estuary, which would have brought them officially under the supervision, and perhaps the tutelage, of Stephen Borough. Indeed, as a ship's master for the Muscovy Company, Pett must have already met with Borough's approval before his selection for such a mission; Jackman probably had some direct or indirect exposure to the navigational instruction of John Dee and the Boroughs through his participation in Frobisher's voyages. Yet despite their impressive backgrounds in maritime exploration, they were subjected to another round of last-minute navigational tutoring.

Unlike Frobisher, Jackman and Pett were undertaking their voyage in the name of the Muscovy Company, whose partners funded the whole venture. The investors were growing impatient and were eager for more positive results this time around, leading the company's governors to take a more active hand in setting out the exact terms of the mission. In their commission to the two seamen, the governors stipulated where they were supposed to explore, suggested when and where they might winter, and ordered them not to get sidetracked exploring every bay and river but to press on to Cathay if it was possible to get there. They placed Pett in command of the expedition, perhaps because he was the more experienced commander, or else because he was an entity better known to them as a Muscovy Company pilot (Jackman's voyages for Frobisher were licensed but not undertaken by the Muscovy Company). The men were ordered to keep their small fleet of two ships together at all times, in order that they might "meet often together, to talk, confer, consult, and agree how, and by what means you may best perform this purposed voyage, according to our intents." They were also told to include their respective mates in all navigational deliberations, "to the end, that whatsoever God should dispose of either of you, yet they may have some instructions and knowledge how to deal in your place, or places."[75]

After telling their highly experienced ships' masters where to go and how to get there, the company put Jackman and Pett under the tutelage of William Borough, who gave them further assistance by providing nautical charts and instructions for their use. Borough's involvement was not as informal on this occasion as it was in the Frobisher voyages; his instructions to Jackman and Pett were written out in triplicate, with one copy given to each of them and one copy "remaining with us, the said company, sealed and subscribed" by both men. The charts Borough provided were unusual because they were "in plat of spiral lines"; this almost certainly meant that they were circumpolar charts, much like those Frobisher had received from John Dee in 1576. The circumpolar projection was potentially much more useful for Arctic voyages than a typical plane chart, because it distorted coastlines the least at high latitudes; but it was also a more specialized and mathematically sophisticated technology, seldom used by the English before 1580. Jackman and Pett would thus very likely have needed Borough's help to understand and use it.[76]

Although Borough's charts were certainly quite advanced by English standards, his brief surviving notes covered mostly basic dead-reckoning techniques, things for which Pett and Jackman would surely have needed no extra instruction. He advised them to be scrupulous about turning the half-hour glass on time, in order to estimate their speed and distance traveled most accurately; to be sure to take depth measurements at least once every two hours, and more often if in shallow water; to record the direction and estimated speed of the ship every two hours, as well as wind direction and any other pertinent course-altering information, such as leeward drift; to take careful note of the time, strength, and direction of tides and currents; and to sketch any prominent coastal features from multiple perspectives, for later use as landmarks. On a more mathematical level, he admonished them to record latitude and compass variation measurements whenever possible, especially if they ever went ashore, together with the time and place of the measurements, and to fill in their observations on the blank charts they carried as often as they could. Beyond that, he even urged them to note the conditions of all the people and lands they encountered along the way, so that the company might trade with them more profitably. In short, Borough told Jackman and Pett to "observe," "draw," and "note diligently," in a comprehensive and permanent record, every scrap of navigational and piloting data they could collect: "*These orders if you diligently observe, you may thereby perfectly set down in the plats that I have given you your whole travel and description of your discovery, which is a thing that will be chiefly expected at your hands.*"[77]

The instructions William Borough provided to Jackman and Pett differed in both character and intent from those given to Frobisher four years earlier. The principal goal in 1576 had been to teach Frobisher and his fellow officers how to navigate in unknown waters using the new mathematical technologies; Jackman and Pett, however, would already have had some experience with this. Borough's emphasis in 1580 was on *recording* the progress of their navigation. In effect, Borough wanted the two explorers to start with the discoveries made by his brother and himself as far as Kholmogory in 1557 and extend them all the way to Cathay, in exactly the same documentary fashion the Boroughs had used. The Muscovy Company expected the pair to return home with nothing less than a reliable chart and rutter by which English pilots could repeat their voyage, just as succeeding ships' masters had learned to follow Stephen Borough's previously charted route. William Borough's instructions were not intended to teach Jackman and Pett how to navigate mathematically, skills that might be assumed of two such experienced mariners and explorers by 1580; rather, he was teaching them how to show the way to others. Borough looked to Pett and Jackman as the next generation of mathematical navigators in England, leading the way for those who would come after them in turn—he was, in essence, passing the navigational torch. For their part, the Muscovy Company gave Borough's advice so much weight in their commission to Pett and Jackman not because they lacked confidence in their mariners' abilities, but because there was so much at stake; if the explorers' mission was successful, the company would profit hugely from their production of a rutter to Cathay, just as they had profited in the Russia trade after 1557.

## Mathematical Navigation and the Role of Expertise

Before 1550 English piloting was not a mathematical art, nor did it appear to be an apt focus for a thorough mathematical reinterpretation, rooted as it was in personal experience and mastery of an endless list of messy local details and incidental circumstances. The changing needs and ambitions of the English mercantile community, however, spurred their merchant marine to set out into unfamiliar seas in search of more profitable and stable markets. This, in turn, fostered their adoption and adaptation of new navigational technologies, developed by Iberian and Italian mariners and cosmographers and based on mathematical and astronomical principles rather than personal experience. This shift in the very foundation of the navigator's art, from the local and circumstantial to the mathematical and universal, liberated English pilots from their familiar coasts and routes and helped them to navigate with some confidence anywhere in the world. Mathematical navigation not only made nautical exploration possible; it also allowed other pilots to retrace the routes of the explorers who discovered them.

The new technologies, once introduced, quickly and permanently altered what it meant to be an English pilot or ship's master. The vast majority of English mariners, it is true, continued to ply their traditional routes in pursuit of long-established markets and so would have had little use for many of the technologies introduced after 1550. But once mathematically minded officers such as Sebastian Cabot and Stephen Borough acquired the authority to train, examine, and select pilots for English ships, the stakes were raised even for those not engaged in exploration. In 1578 George Best reported, with perhaps a trace of bravado, that "[t]he making and pricking of cards, the shifting of sun and moon, the use of the compass, the hour-glass for observing time, instruments of astronomy to take longitudes and latitudes of countries, and many other helps, are so commonly known of every mariner nowadays, that he that hath been twice at sea, is ashamed to come home, if he be not able to render account of all these particularities."[78] William Bourne, the author of a very popular English navigation manual entitled *A Regiment for the Sea*, confirmed Best's observation. In the preface to the third edition of his manual, printed in 1580, Bourne described how dramatically the English art of piloting had changed, with the new mathematical technologies rising from derision to dominance: "I have known within these 20 years that them that were ancient masters of ships hath derided and mocked them that have occupied their cards and plats, and also the observation of the altitude of the Pole [Star]. . . . Wherefore now judge of their skills, considering that these two points is the principal matters in navigation."[79]

Yet the reasons for the rapid triumph of the new mathematical technologies are far from obvious—their value as navigational tools could not have been self-evident to a conservative community of practitioners who had prospered without them for decades. Before their merits were demonstrated in actual practice, there can have been little empirical reason for experienced but uneducated pilots to leave behind the traditional methods they knew and had long depended upon in favor of tricky, confusing, and often expensive innovations. How were English pilots persuaded to alter or augment their old art and adopt such a thorough and professionally threatening reinterpretation of it?

Mathematical technologies of navigation could never have become so dominant in England as quickly as they did without the enthusiastic advocacy of two key individuals: Sebastian Cabot and Stephen Borough. Both men were well respected in England, by practitioners and patrons alike, as masters of their art, and their reputations as experts won them considerable administrative authority as overseers of English pilots and teachers of the new methods. They had both the means to foster and the power to demand the adoption of mathematical innovations by English pilots throughout much of the Royal Navy and the merchant marine. Yet their command of the new technologies was not in and of itself the source of their compelling authority. Once again, the practical value of their mathematical skills could not simply be assumed but had to be established in an English context that was alien and potentially hostile to it. The authority of Cabot and Borough as navigational experts was based instead upon two criteria: their prior success as explorers and their association with a renowned and respected navigational institution, the Spanish Casa de Contratación.

On the one hand, as successful (and perhaps more to the point, surviving) sixteenth-century maritime explorers, Cabot and Borough had a heightened credibility among the practicing pilots who were the target audience for the lessons they had to teach. Because they were seamen themselves, they understood the experiences, needs, anxieties, and limitations of their pupils. Their connection with the Casa de Contratación, on the other hand, gave them an enhanced credibility with English mercantile investors, who hoped to build a global trading empire for England on the Spanish model. English merchants, and the privy councillors who supported them, perceived that part of Spain's enormous success lay in its mastery of the seas—the ability to sail dependably across the world from one market to another. The Casa helped make this possible, they believed, by generating a corps of well-trained Spanish pilots on the cutting edge of their art. To compete with Spain, therefore, England had to develop its own version of the Casa, and English patrons naturally turned to the men who had firsthand knowledge of its workings to duplicate its function. Cabot and Borough, then, possessed a dual authority, with both patrons and practitioners; their expertise allowed them to serve as mediators between the two groups, and they were therefore well positioned to reshape the art of navigation in England in both circles, according to their own beliefs and priorities.

The mathematization of English navigation serves as a particularly striking example of a broader phenomenon. As the notion of technical expertise evolved during the sixteenth century, its emphasis shifted from a basis in experience to the possession of a body of abstracted and generalized knowledge. Whereas expertise had once been rooted in mastery of the local and the particular, it gradually came to be portable and universal, ostensibly applicable to problems and circumstances that had never been encountered before. Moreover, the generalized principles of a given field of expertise could be codified and communicated to others, stretching beyond the limited experience of a single practitioner to become a publicly shared commodity, through the writing and publication of instructional manuals. Just as a mathematical navigator could leave his familiar coasts and practice his art from any point on Earth, so expertise was set loose from its local, empirical moorings to be read and admired across Europe.

In the process, the experts themselves grew increasingly distinct from mere practitioners; their possession of a special kind of theoretical knowledge placed them on a new and elevated level. In distancing themselves from the actual manual labor of the art in question and emphasizing their deeper understanding of its basic principles, experts attained a higher intellectual and social status. They enjoyed more direct and extended contact with wealthy and powerful patrons and assumed positions of authority over their practitioner colleagues. The experts ultimately came to serve as mediators, moving freely between patrons and practitioners, acting as managers of the latter on behalf of the former. As their patrons came to trust and rely upon them more fully, the experts grew in number; they began to be aware of the power they held and to wield it self-consciously. In chapter 4 we move from the Arctic seas to the lecture halls, private studies, and printers' shops of England, from which mathematicians boldly proclaimed their expertise in an art that most had never actually practiced.

CHAPTER FOUR

# Secants, Sailors, and Elizabethan Manuals of Navigation

> [A]lthough my chiefest intent hath been to pleasure those that shall have occasion to put the thing in practice by their own travail and experience, yet because some of the rules are deducted from the fountains of the mathematical sciences, and wrought by the doctrine of sines and triangles, which may seem strange in our English tongue, and wherewith few seamen are yet acquainted, I may seem to have missed of my first good meaning.
> WILLIAM BOROUGH, *A Discovrs of the Variation of the Cumpas*

> [T]he true understanding and reason of the nautical planisphere or sea-chart, may by him that hath been but meanly conversant in mathematical meditations be better apprehended, than otherwise it can by the seafaring man, though he spend his whole life in sailing over all the seas in the world.
> EDWARD WRIGHT, *Certaine Errors in Navigation*

In his well-known "Mathematicall Praeface" to Henry Billingsley's 1570 English translation of Euclid, the mathematician John Dee expounded at length on the boundless practical utility of the "sciences, and arts mathematical," among which he included the art of navigation. Navigation, Dee wrote, drew liberally upon the other mathematical arts he had already discussed, including hydrography, astronomy, astrology, and horometry. These arts, in turn, all derived ultimately from "the common base, and foundation of all: namely *arithmetic* and *geometry*."[1] A thorough knowledge of mathematics was thus necessary, Dee argued, if the navigator hoped to understand the proper construction and use of the myriad mathematical instruments that were rapidly becoming requisite for his art, such as the cross-staff, the astrolabe, the plane chart, and various astronomical tables.

Up to this point, thanks largely to the pedagogical efforts of Sebastian Cabot and Stephen Borough, a growing number of practicing English pilots and ships' masters might well have concurred. But John Dee, mathematician and astrologer to Queen Elizabeth and her court, was advocating no mere practical application of mathematics; the recommendations contained in his "Praeface" called for mathematical *mastery* in the art of navigation. He referred not simply to the cross-staff, but specifically to the "astronomer's staff," calling attention to its mathematical origins and lineage. He advocated using not the basic, stripped-down mariner's astrolabe, but nothing less than the full astronomer's "astrolabe universal," though most of its extra functions would have been of very little use at sea. He endorsed not the common plane chart, then widely used throughout Europe, but what he called the "true" variety, one "not with parallel meridians," possibly referring to his own, still obscure, circumpolar projection. And according to Dee, the truly competent navigator should not merely know how to use astronomical tables but also be able to calculate his own. The navigator, in short, was to be in total command of his mathematical art, so that he might "understand, and judge his own necessary instruments, and furniture necessary: whether they be perfectly made or no; and also can (if need be) make them himself."[2]

Dee's conception of the navigator's art, by all rights, might have seemed unreasonable to many of his readers. Dee was calling for the mathematization of navigation on a level that far surpassed the practical needs and mathematical capabilities of virtually all Elizabethan pilots and ships' masters, many of whom were still neither literate nor numerate. This is less surprising, perhaps, when one considers that he was hardly approaching the subject from the mariners' perspective. After all, Dee himself had no real nautical experience—other than a few Channel crossings as a passenger, he had never even been to sea. Yet Dee's vision of the art of navigation, unreasonable or otherwise, was certainly not unique in Elizabethan England, nor was it the most ambitiously or radically mathematical. During the last quarter of the sixteenth century, a great many English treatises on navigation were published, introducing new technologies based on ever more sophisticated mathematics. The treatises' level of mathematical content rapidly overwhelmed the meager training of most contemporary practicing navigators, such as Frobisher, Jackman, and Pett. Nevertheless, these books were popular; many of them went through several editions in the space of only a few decades. By 1600 the art of navigation, as presented in the host of English manuals purporting to teach it, was so dominated by mathematical technologies that it came to be wholly subsumed as a branch of applied mathematics. In the process, mathematicians gained an even greater authority over the discipline than its practitioners possessed: to qualify as a navigational expert in 1600, a knowledge of trigonometry was more important in some circles than actual experience at sea, and mathematicians became the masters of an art that many had never even seen in action, let alone practiced themselves.

## The New Masters of Navigation

This chapter continues to explore the development of mathematical navigation in England through the end of the sixteenth century, but its protagonists as a group are very different from those of the previous chapter. Sebastian Cabot and Stephen Borough were, above all else, experienced mariners, trying to reach an audience of others like themselves. Their authority as navigators in England was based largely upon their prior successes in practicing the art they sought to teach and improve. The real maritime experience of the authors of mathematical navigation manuals, in contrast, was often negligible. Several of them, like Dee, had rarely if ever been to sea; their authority as experts in navigation was based instead upon their superior knowledge of mathematics, especially certain applications of geometry, trigonometry, and astronomy. By the end of the century, the art they expounded in their manuals bore little resemblance to anything taught by Cabot and Borough to practicing pilots during the 1550s and 1560s, and even less to what those pilots actually did in guiding their ships at sea.

The question to be asked here is Cui bono? How and why did navigation come to be so thoroughly mathematical? Whose interests drove this fundamental transformation of a discipline? Most manuals' nominal audience, and by implication their intended beneficiaries, were the English mariners who desired to improve their knowledge and skill, to attain a better command over their rapidly changing art. Some practicing navigators may well have read a few of the manuals, and perhaps they even profited from them; some seamen were more mathematically inclined than others, of course, and some of the manuals were less strictly mathematical and more practically oriented. Yet in the late sixteenth century, navigation was still an art learned not from books but through long experience at sea, increasingly organized around a formal apprenticeship and with little opportunity to obtain a rigorous mathematical education.[3] By 1575 even such relatively simple equipment as the cross-staff, the mariner's astrolabe, and the plane chart had been used aboard English ships for just a decade or two at most; only a tiny handful of English seamen could yet handle more sophisticated technologies such as the compass of variation or the circumpolar projection. Indeed, the vast majority of English mariners still plied only their old, familiar routes, having little if any need for the latest mathematical innovations. It could not have been Elizabethan seamen alone, for most of whom long division still posed a considerable challenge, who were consuming the thousands of mathematical volumes supposedly printed for their benefit. The prefaces and general tone of the manuals themselves often indicate that their principal intended audience was actually a far less nautical one: the mercantile, gentry, and noble classes of Elizabethan England.

The early modern period saw a surge in demand among royal administrators and courtly patrons for clients with mathematical expertise, not only in England but throughout western Europe. From Spanish cartography to Italian ballistics to German astronomy, historians have repeatedly shown how the mathematical arts came to be ever more established in Europe's princely courts.[4] In each case, mathematicians appeared to offer their ruling-class patrons new solutions to pressing problems, giving them greater control over their increasingly powerful and centralized nation-states. The desire for mathematical expertise was especially intense in England, where the humanist-inspired patronage networks were so heavily focused on action-oriented knowledge and practical results.[5] This trend toward mathematical patronage can be traced back to the reign of Queen Elizabeth's father, Henry VIII. During the 1530s and 1540s, anticipating an invasion from the Continent in response to England's split from the Catholic church, Henry eagerly sought to modernize his coastal defenses against the threat of cannon fire, by then a principal means of warfare in northern Europe. Self-consciously following Italian and Dutch examples, he brought Continental masters of the newly mathematized arts of fortification and gunnery to England and made sure that native Englishmen soon acquired their expertise. At the same time, as the dissolution of the English monasteries created an unprecedented boom in land speculation, surveying and cartography were in high demand, and these arts were similarly reconceived as practical applications of geometry and trigonometry.[6]

The sudden yet enormous relevance of all these newly mathematical arts to the Tudor ruling classes created a pressing need among them for at least a passing familiarity with mathematics. Beginning in Henry's reign, English gentlemen and aristocrats commonly spent some of their youth at university, where many were exposed to mathematical studies.[7] Some wealthy families even employed mathematical tutors in their households so that their children might get an early start— John Dee served in such a capacity in the household of the powerful Duke of Northumberland during the reign of Edward VI, for example. But basic mathematical proficiency was a necessity not limited to the landowning classes. Below the ranks of the nobility and gentry, English merchants also possessed a significant degree of mathematical training and acquired considerable skill in practical mathematics in the course of their everyday business dealings.

More than any other mathematical art, navigation was a common and pressing concern among members of all these groups. Besides the Crown's obvious concern for the military effectiveness of the Royal Navy, the operative range and well-being of the English merchant marine was vital to the interests of scores of wealthy and powerful investors, including royal courtiers and administrators as well as merchants. As shown in the previous chapter, oceanic voyages of exploration and trade once again became an important component of the English mercantile economy after 1550. These voyages could yield incredible profits when successful, but their failure could mean staggering losses for those who had invested in them; so it was crucial to maximize the chances for success. The investors thus had a considerable stake and natural interest in keeping abreast of the latest navigational technologies. The ability to navigate efficiently and dependably was of paramount importance, after all, if England hoped to compete and prosper as a maritime power, especially against its formidable Iberian rivals. Members of England's wealthy elite thus formed the principal audience for treatises addressing navigation, and the mathematical arts more generally.

Nevertheless, the principal beneficiaries of the dozens of mathematical manuals written during the sixteenth century were not the investors and men of leisure who bought them, but the authors who wrote them as a means of winning patronage. As demand for their valuable knowledge grew, mathematicians became an ever more active and visible conglomerate in England. Those seeking advancement might make elaborate gifts of mathematical instruments they had constructed or treatises they had written, and they were usually well rewarded for their efforts by the patrons they cultivated. The growing importance of mathematics to English patrons made it an ideal route for a skilled client to become an expert mediator. Mathematicians could appeal directly to their patrons' demand for useful, active knowledge on the part of their clients. They based their appeal upon their mastery of a body of abstract principles and theories, by means of which they had brought several practical and valuable arts within their province. They gained control of these arts by recasting them in their own image, reinterpreting the basic tenets and problems in mathematical terms, and providing a series of new mathematical technologies to address them, which neither practitioners nor patrons could comprehend without their assistance. They presented the new technologies as being inherently more valuable than traditional craft practices, because they were based not on limited practical experience, but on general theoretical principles at the root of that experience, making the mathematical approach appear to be more widely applicable and less ad hoc.

Once patrons and practitioners alike had been made to accept the mathematicians' reconception of their traditional arts, further progress and development became impossible without recourse to the mathematical experts. A new and powerful stratum of expert mediators was thus created, which dominated the mere practitioner on the one hand and served as the sole source of expertise to its patrons on the other. In short, mathematicians vastly improved their position by making mathematics an *obligatory point of passage* for anyone hoping to understand or control the arts in question.[8] So successful were mathematicians in co-opting the art of navigation, in particular, that by 1600 no English pilot could be considered an expert navigator without a knowledge of arithmetic, geometry, and trigonometry, whereas mathematicians who had never been to sea could portray themselves as the only source for true navigational expertise.

In order to lay claim to their newfound power and status, the mathematicians first had to make their expertise public and advertise the valuable services they could offer their patrons. The best way to accomplish this was to publish manuals and treatises, promising to make the mathematical arts more accessible and beneficial to anyone who would accept them on the mathematicians' terms. The navigation manuals considered here were just one part of a consistent program of English mathematization. They formed a prominent subgenre within a much larger body of works dedicated to the study of the mathematical arts, which also included (among others) surveying, mensuration, gunnery, fortification, and military science. These books were pedagogical in style, nearly always in the vernacular, and usually aimed at a less mathematically sophisticated audience. Taken together, they helped to define the mathematical arts themselves and to create a new stratum of mathematical experts who shared many of the same tools, techniques, and ways of approaching a broad spectrum of problems. They extolled the study of mathematics as both a valuable aid to business and a pleasant, mind-strengthening recreation for the "better sort," even as they ignored the mundane, practical needs and educational limitations of actual practitioners. As the mathematical authors spread their influence across a number of vital and profitable pursuits, they presented themselves to their patrons as ever more indispensable assets of the commonwealth, raising their own intellectual and social status accordingly.

The growth and proliferation of the mathematical arts in early modern England has often been interpreted through the emergence of a new intellectual and cultural community, usually referred to as "mathematical practitioners."[9] This community has been defined as possessing not only a shared knowledge of and devotion to practical mathematics but also a common mathematical perception of the world around them. Although the increasing prominence of the mathematical arts in Elizabethan England cannot be denied, ideas regarding the formation of a coherent community within and around them must be critically reexamined. In her comprehensive census of mathematical practitioners in Tudor and early Stuart England, for example, E. G. R. Taylor has encompassed quite a disparate group of individuals, including gentlemen, courtiers, academic mathematicians, physicians, astrologers, instrument makers, mariners, gunners, surveyors, and merchants, among others. These men certainly shared some engagement with practical mathematics and were often in contact with one another across their various social boundaries and diverse backgrounds. Yet it may well be doubted whether they themselves would have, or even could have, self-identified as any sort of community; and therefore the question arises whether it is most useful or informative for historians to treat them as one.

Considering the members of a given group as a single, coherent community can sometimes help historians to envision and understand more clearly their common interests and activities, but it can also obscure important interactions *within* the group in question. Rather than focusing exclusively on the group as a whole, the particular contributions, responsibilities, and interests of each individual must be examined, or the complex relations between the various members of a putative community might be misunderstood or even effaced altogether. Although many Englishmen did share some knowledge of mathematics and a desire to apply it somehow toward practical ends, by itself this should not lead historians to view wealthy merchants, university mathematicians, artisanal instrument makers, and practicing seamen as some sort of mathematical fraternity with common goals and interests, even if they did have occasional professional contact with one another. Beyond their disparate backgrounds, their various mathematical contributions and social interests were often sharply at odds with one another, and bitter rivalries sometimes broke out between them.[10]

Far from helping to form a community, the mathematization of certain practical arts in early modern England, and especially navigation, actually helped to *divide* one. In order to gain the patronage they desired, proponents of the mathematical arts first had to differentiate themselves from the mass of common practitioners. As these mathematically minded clients worked to identify themselves with the cultured and learned, became servants of the wealthy and powerful, and even began to move in courtly circles, they created a social and intellectual fissure between themselves and actual practitioners, whom they regarded and often portrayed as mere unlearned craftsmen.[11] In serving as expert mediators, they exploited their superior command of mathematics to separate patrons and practitioners more widely than ever before. The same point may be made about expert mediators in general, even without the mathematization of the field in question; the mediating efforts of the expert often helped to create or reinforce boundaries between groups, even as the experts overtly sought to bridge them. Thomas Digges's mediation during the rebuilding of Dover Harbor, for example, helped to connect patrons and practitioners more productively and to build a consensus between them toward a common goal. Yet at the same time, his approach served to keep patrons and practitioners physically and administratively distant from one another throughout most of the project, while elevating him above his fellow Dover commissioners and the Romney craftsmen who actually carried out the work.

This chapter examines in detail a number of mathematical manuals on the art of navigation. It cannot be, and is not intended to be, an exhaustive treatment; rather, a series of examples are considered from works written throughout the last quarter of the sixteenth century, by authors of diverse backgrounds, in order to convey a general impression of the genre as whole. For each example, attention is paid to the author, his background, his approach to the subject, his intended audience, and the prior knowledge and skill he assumed on the audience's part. The style and approach of each manual is considered especially with respect to a particular problem—using the sun's elevation and declination to determine one's latitude—so that comparisons may be drawn between the manuals. The authors examined here co-opted and redefined the art of navigation over time to suit their own patronage needs. They aimed their mathematical interpretations of navigation primarily at an audience of patrons with an amateur's interest in the art, in order to elevate themselves intellectually and socially at the expense of the mariners who actually practiced it.

## Navigation without Ships: The Navigational Manuals

### *William Bourne*, A Regiment for the Sea

Richard Eden's 1561 translation of Martín Cortés's work *The Arte of Nauigation* was the first navigation manual printed in England, and it introduced its English readers to a number of new mathematical technologies adapted for use at sea. In 1574 the Englishman William Bourne published his own manual, entitled *A Regiment for the Sea*, supplementing Cortés's book by filling in some of its most glaring lacunae. Continuing the Iberian drive to mathematize navigation, Bourne's manual was one of the most influential of the sixteenth century, as well as one of the most popular; it had a second, unauthorized edition in 1577, prompting an authorized, corrected edition in 1580. There followed at least eight more English editions, the last in 1631, as well as three editions in Dutch.[12] Because of its early date of publication, its popularity over many decades, and its comprehensive yet comparatively basic treatment of mathematical navigation technologies, *A Regiment for the Sea* acted as a sort of baseline for many later manuals. It provided the foundation of navigational knowledge that would later be assumed of the interested reader and upon which succeeding treatises attempted to build.

Bourne was a native of the Thames port town of Gravesend, just downriver from London. A self-educated man, he was nevertheless skilled in many mathematical arts and wrote treatises on mensuration, surveying, and gunnery, as well as navigation. Despite his interest in maritime practice, Bourne was never a seaman by trade. He apparently made his living as an innkeeper, surveyor, and shore gunner and also served in various local administrative posts such as jurat and port reeve. Nevertheless, his activities in and around the busy harbor at Gravesend would have given him ample opportunity to confer with local seamen and pilots, whereas his service as a shore gunner probably explains his connections to the Royal Navy. He dedicated his *Regiment* to the lord high admiral, the Earl of Lincoln, and another treatise, called *The Treasure for Travellers*, was dedicated to William Wynter, the navy's surveyor and master of ordnance.

In his preface to *Regiment*, Bourne proclaimed his intention of writing for common, practicing mariners who wished to improve upon their knowledge and skill. Despite Stephen Borough's years of instruction, according to Bourne, the average English seaman of the early 1570s did not even know the names of the tools required for his art, let alone their proper use. Far from trying to write on the level "of excellent learned men in the mathematical science," therefore, Bourne sought to make the fruits of that science available to a wider audience, so that no one might be left behind: "And albeit the learned sort of seafaring men have no need of this book, yet am I assured that it is a necessary book for the simplest sort of seafaring men."[13] The lack of any formal education on the part of such readers was to be no excuse for their ignorance of the basic technologies of mathematical navigation, for Bourne described himself as an "utterly unlearned" man, who had taught himself mathematics and mastered the new skills "without help of any learned persons."[14]

Although Bourne addressed many of the same subjects as Cortés, he generally did so in greater detail, with superior organization and clarity. His instructions for actually using the various navigational instruments of the day were a strong improvement over Cortés's, whose treatment of the same was often vague, incomplete, or opaque.[15] For example, Cortés's explanation of how to use the sun's altitude and declination to calculate one's latitude was disastrously confusing. This calculation, though straightforward in principle and of great utility, could nevertheless be a remarkably complicated affair. Although only the addition and subtraction of angular measures was actually required, the algorithm that told the navigator when to perform which operation and with which set of data could be tricky to follow and even trickier to explain to others clearly. The exact procedure to be used depended upon two variables: the date (or more precisely, the sun's position along the ecliptic) and the relative position of the ship with respect to both the equator and the sun's current declination.[16]

Cortés's treatment of the subject began with a brief consideration of the astronomical and geometric concepts involved. Though necessary in order for the rest of the lesson to make any sense to the reader, his explanation of the basic terms he used seems so cursory as to be almost incomprehensible without a prior knowledge of spherical geometry and solar astronomy:

> It is convenient to define the altitude before we give rules of the use thereof. The altitude of the sun or the moon, or of any other star, is the distance that is between it and the horizon. And this ought to be accounted by the degrees of the greater circle which passeth by the zenith and by the center of the sun or of the moon, or of the star unto the horizon. And the degrees that are from the horizon to the star or to the sun, that is the altitude; and the degrees that are from the center of the star or of the sun unto the zenith, is called the complement or supplement of the altitude. The altitude of the equinoctial [celestial equator] is ever counted by the meridian. And the degrees of the meridian that are between the equinoctial and the horizon, is the altitude of the equinoctial. And other so many, are they that are from the zenith to the pole. For the altitude of the equinoctial, is equal to the complement of the altitude of the pole. The degrees of the meridian that are between the equinoctial and the zenith, is called the complement of the altitude of the equinoctial: and is equal to the altitude of the pole. And although we have defined the altitude in general, yet shall we only profit ourselves by the meridional altitude of the sun. The meridian altitude, is the greatest altitude that the sun hath everyday. And this shall be when the center of the sun is in the meridian. And the arc of the meridian that is between the horizon and the sun, is the meridian altitude. So that when we say the altitude of the sun is taken, it is understood at mid-day. The shadows that the sun then maketh are in three sorts. For either to us it casteth the shadow toward the north part, or toward the south, or perpendicular by a right up line, so that at midday or noon, nothing that standeth upright, giveth any shadow at all.[17]

Cortés followed this summary overview with an abrupt explanation as to how the reader was to render his calculation, under every conceivable circumstance:

> But for as much as there is such variation in declinations, altitudes, shadows, and parallels, it shall be necessary to give rules for all variations. And these shall be reduced into four brief and compendious rules: the which I have here described that

the witty may take profit by them, and the rude learn them, not caring for the rules of the mariners, because they are too long and tedious. For (as the Philosopher sayeth) it is vainly done by many, that may well be done by few.

When the shadow shall be perpendicular, it is because the sun is in the zenith, and 90 degrees above the horizon. And then how many degrees of declination the sun hath, so much shall we be distant from the equinoctial toward the part where the sun declineth. And if it have no declination, it and we shall be under the equinoctial.

When the sun and the shadows shall be to us from the equinoctial toward one of the poles, we shall take away the declination from the meridian altitude. And the complement for 90 shall we be distant from the equinoctial toward the same pole.

When the sun declineth from the equinoctial toward the one pole, and the shadows shall be toward the other, we shall join the declination with the meridian altitude. And if all come not to 90 then the complement for the 90 shall we be distant from the equinoctial toward that pole to the which the shadow falleth. And if they be more in number than 90 then the overplus of 90 shall we be distant from the equinoctial toward the pole where the sun declineth. And if they be just 90 we shall be under the equinoctial.

When the sun hath no declination, we shall be distant from the equinoctial the complement of the meridian altitude toward the pole where the shadows are.[18]

In the remainder of the chapter, Cortés provided his readers with three "examples" to aid their understanding, but ironically, none of these would have been of any help to a seaman trying to find his latitude at sea using these rules.[19] In fact, Cortés gave minimal practical assistance to his readers, including not a single sample calculation that a reader-practitioner of limited education might use as a working model. Despite the long list of varying circumstances under which his procedure might be applied, therefore, Cortés's emphasis was on articulating the underlying theory fully, perhaps for readers who already had some grasp of the basic principles, rather than on helping unlearned pilots put it to actual use at sea. This might have been all right in the Spanish context for which the book was written—Cortés, after all, could assume that his Spanish readers would have had some personal instruction at the Casa de Contratación to supplement his lessons—but for most English seamen the book would have been practically worthless. While some of "the witty" might have managed to profit from reading Cortés's manual, "the rude" were surely left to fend for themselves.

William Bourne's approach was entirely different; to begin with, his discussion of the same topic filled eight leaves, compared to Cortés's two.[20] This alone allowed the presentation of a less hurried and more detailed treatment. He also did not emphasize the astronomical foundations of the procedure; indeed, he had even less to say about astronomy than Cortés did, despite his lengthier discussion. However, Bourne crafted his instructions in such a way that no real understanding of astronomy was necessary; his readers could still use the mathematical technique he was demonstrating, even if they did not fully understand how and why it worked, so long as they followed closely the method he described. He accomplished this by breaking down the complex topic into several chapters, one for each set of relevant circumstances, and providing copious examples.[21] The reader had only to determine which chapter and set of examples most closely matched his current situation and substitute his own measurements for the numbers in Bourne's sample calculations. In order to select the appropriate chapter, the reader only needed to know the date and have a rough idea of how near his ship was to the equator, relative to the sun. The date was used to look up the solar declination in the detailed table that Bourne provided in chapter 5 of his book; Cortés had provided a similar table, but Bourne's was expanded and reorganized so that it too was simpler to use.[22] Finally, to better illustrate the whole process, Bourne added schematic diagrams to each example, allowing the reader to form some visual/geometric understanding of the problem.

To take but one example, chapter 8 of Bourne's text (the most complicated of those addressing this problem) explained how to find one's latitude when sailing near the equator:

> Now furthermore if you be unto the south parts near unto the equinoctial, so that the sun have any great declination either to the southwards or the northwards, you being between the equinoctial and the sun, when you have taken the true height of the sun with the astrolabe, to know the height of any of the 2 poles, do this: Seek the declination of the sun for that day with the degrees and minutes; the declination being known and the height of the sun in like manner, then add the declination of the sun unto the height thereof, and it will exceed or be more than 90 degrees; then again look how many degrees it is more than 90 with degrees and minutes, that shall be the true height of the pole towards that side that the sun is, because the equinoctial is the number of degrees above 90 (which is your zenith) to the contrary part from the sunwards. For (as I have said in the chapter going before, and is general forever) look what height soever the equinoctial be from the horizon, that is the true distance between the zenith and the pole; in like manner look what distance is between the equinoctial and the zenith, the same is the true distance between the horizon and the pole, that is to say, the pole is so many degrees in altitude above the horizon. As it is a common saying (in knowing how far we be unto the southwards or northwards) that

the pole Arctic is so many degrees in altitude, or (as some will say) that we are in so many degrees in latitude: the question is all one in effect, although the one be called altitude or height, and the other latitude or wideness, yet it hath one signification. For as when you say altitude or height of the pole, you mean the pole is raised so many degrees above the horizon, so likewise when you say latitude, you mean you be so many degrees in wideness from the equinoctial: for that your zenith or vertical point is so many degrees from the equinoct. Moreover if you chance to be right under the equinoctial, as you cannot say that you have any latitude, so likewise cannot you say that you have any altitude, for that the two poles be then just with your horizon, and in like manner the equinoctial is your zenith or vertical point. But when you will take the height of the sun with your astrolabe, then look what declination the sun hath, either to the southwards or northwards. Then put the declination of the sun unto the height of the same, and the number will be just 90 degrees: if it lacketh anything of 90 degrees, then it signifieth that the equinoctial lacketh so much of the zenith, and so much just shall the pole be above the horizon towards that part that you be in from the sunwards. But contrariwise, if it doth exceed or be anything more than 90 degrees, then (as afore is declared) it signifieth that the equinoctial is as much as that number (both in degrees and minutes) on the contrary side from the sunwards, that is to say, your zenith shall be between the sun and the equinoctial, and the pole shall be so many degrees or minutes above the horizon, as is the distance between the zenith and the equinoctial, towards that part or side that the sun is on.[23]

Bourne obviously found it at least as difficult as Cortés had to explain the nuances of this tricky concept clearly and concisely in prose, though the reader's confusion might have been somewhat mitigated by the much greater care and attention Bourne used in treating the subject at such length. In any case, Bourne then went on to clarify his convoluted explanation through a series of concrete examples, with accompanying diagrams:

Wherefore I do think it necessary to give certain examples (and first take this for an example). Admit I do take the height of the sun unto the northwards 80 degrees above the horizon, and the sun hath declination unto the northwards 20 degrees, to which I add or put the height, that is to say 80 degrees (being the height of the sun) and 20 degrees (being the declination of the sun) do make 100, from which I pull 90 away (which is my zenith) and so there remaineth 10 degrees. Wherefore you may conclude, that the equinoctial is 10 degrees to the south part of your zenith, and the sun to be 10 degrees to the north part of your zenith, so that the north pole is 10 degrees above the horizon, as by this example it is declared.

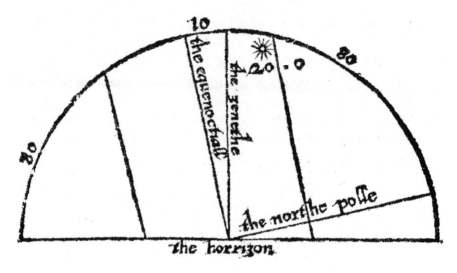

And for the second example, admit I take the sun unto the northwards 75 degrees and 20 minutes above the horizon, the sun having north declination 14 degrees 40 minutes, I then do add or put 14 degrees 40 minutes unto 75 degrees 20 minutes, and those 2 joined together maketh 90 degrees, whereof you may conclude that the equinoctial is your zenith, and then the 2 poles be with your horizon, as by this example it doth appear.

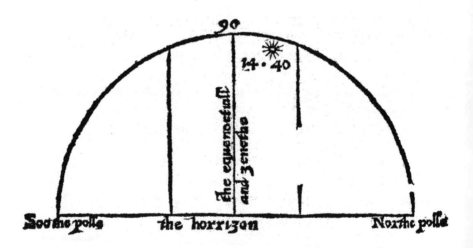

And now followeth the 3[rd] example. I admit the sun be taken with the astrolabe 81 degrees and 15 minutes above the horizon, and the same hath the south declination 22 degrees 35 minutes, wherefore I do add or put together 81 degrees and 15 minutes (being the height of the sun) and 22 degrees 35 minutes (being the declination) and that maketh 103 degrees 50 minutes: from which I take away 90 degrees (which is my zenith) so that there remaineth 13 degrees 50 minutes: so that you may safely conclude, that the equinoctial is 13 degrees 50 minutes unto the north parts of the zenith, and then it must needs follow that the south pole is 13 degrees 50 minutes above the horizon, as by this example it is declared.[24]

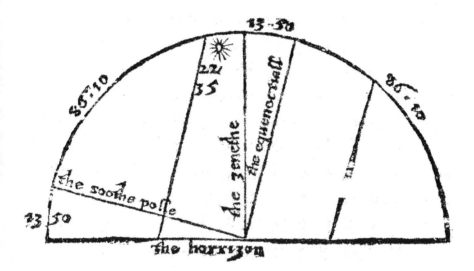

Although Bourne's long explanation of the geometry and astronomy behind his method might not have been much clearer than Cortés's, it was also not strictly necessary for using the technique *as he presented it*. The reader could skip directly from Bourne's instructions to his examples, substituting his own observations for those of the model calculations, and then stick closely to the steps Bourne followed to obtain an answer. He could thus apply the whole process successfully without fully understanding *why* it worked.

This was a strategy that Bourne and other authors used repeatedly in their treatises to deal with the dilemma presented by their readers' relative lack of mathematical ability. The new technologies discussed in the navigation manuals relied upon ever more complicated mathematics and ever more difficult or tedious calculations. The vast majority of readers, however, were unlikely to keep up with the mathematics required to take advantage of them. To solve this problem, the authors of navigation manuals worked to remove the heaviest mathematical responsibility from their readers, taking it upon themselves instead. They completed as much of the necessary calculation as they could while compiling their texts and then presented their readers with a simplified, user-friendly manifestation of the technology in question—an instrument, a table, a model calculation, or perhaps just a basic, fixed method for the reader to follow. These tools allowed the reader to benefit from the new technology even if he did not understand all of the mathematics involved in it, because the bulk of the calculation had already been tackled by the author. The reader could thus concentrate on input and output—observations, measurements, and other such data—and not concern himself with the troublesome mathematics in between. In a word, such tools may be seen as a sixteenth-century version of a *black box*, a device intended to shield the user from complex ideas and processes he did not need to comprehend by making it possible for him to function in a limited input-output role. By deliberately black-boxing their new technologies, the mathematicians made them accessible to a much wider audience. Yet through this very process, they also transformed themselves into an obligatory point of passage for anyone who wished to understand fully, let alone improve upon, the newly mathematized art of navigation. Only the mathematicians retained complete control over the technologies they created and circulated.

Bourne, for example, employed this black-boxing strategy throughout his manual by creating and including in it several tables and other textual instruments. These tools removed the need for computation from the reader, reducing his mathematical responsibilities by relocating them with the author. Bourne's device for finding the linear distance in one degree of longitude at any given latitude is a striking case in point. The original problem is a basic trigonometric calculation—the linear length of one degree of longitude is proportionate to the cosine of the latitude in question. Bourne represented this calculation geometrically in his text, with the latitude and longitude represented on a scaled semicircle. However, in order to apply the method, the reader did not need to know any geometry or trigonometry whatsoever; he had only to add a piece of thread to Bourne's semicircular diagram and stretch it to the appropriate latitude measurement along the circumference of the semicircle. The actual mathematics of the calculation had been effaced from the process by Bourne and replaced with simple instructions for using his textual instrument. Promoting practical utility, rather than the mathematical understanding of the reader, was his highest priority.

Bourne did not forsake mathematical explanation altogether, however; ideally, he wrote, all navigators should have a sound theoretical understanding of what they were doing. He argued that success without reason, without some grasp of the *why* as well as the *how*, might be dismissed as mere good luck: "[W]hatsoever he be that

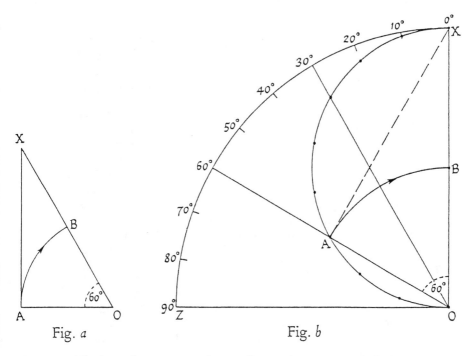

*Figure 4.1.* The linear distance in one degree of longitude at any given latitude is proportional to the cosine of that latitude. Geometrically, if the linear distance in a degree of latitude (a constant) is represented as the diameter of a semicircle (OX), then the length of a chord drawn from one end of the diameter to the circumference of the semicircle (OA) represents the cosine of angle XOA and thus the linear distance in a degree of longitude at that latitude (OB, as a proportion of length OX). This figure is adapted from *A Regiment for the Sea, and other writings on navigation*, by William Bourne of Gravesend, a Gunner, ed. E. G. R. Taylor (Cambridge: Cambridge University Press, for the Hakluyt Society, 1963), 129.

will say that he can do any thing, and if that he cannot show the reason of the doing thereof I do say unto you he cannot do it, and this is most certain for if that he doth it, he doth it but by fortune."[25] To be a true master of the art of navigation, then, was to be fully comfortable with the mathematical theory underlying maritime practice. For this reason, Bourne rarely failed to explain how a given tool or table was created, even when his explanation was not strictly necessary for its use. He also sometimes included extra information that he knew to be of no practical use to the mariner, though in such cases he recommended to his reader-practitioners that they simply ignore it, since it was unnecessary for them: "I would not wish the common mariners to trouble themselves with these matters, but follow their accustomed order."[26]

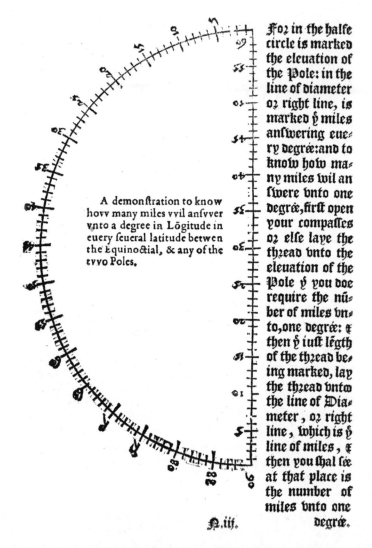

*Figure 4.2.* Bourne's textual instrument for determining the linear distance in one degree of longitude at any given latitude. To use the instrument, the reader had only to stretch a piece of thread from the bottom corner of the diagram to the edge of the semicircle, at whatever latitude measurement he was interested in, marking the length of the thread as he did so. He then placed that length of thread along the semicircle's diameter, which included a scale of miles, from zero to 62. Whatever number the thread reached told him how many miles were in one degree of longitude at that latitude. At 0°, the answer would be 62 miles, the same as in any degree of latitude. As the latitude increased, the thread would grow shorter, and the number of miles would diminish accordingly as the meridians converged toward the poles. To use the instrument successfully, the reader needed no mathematical skills whatsoever; it was a question not of calculation, but of simply measuring the length of the thread. William Bourne, *A Regiment for the Sea* . . . , 3d ed. (London, 1580), fol. 51 r. Photo courtesy of the Beinecke Rare Book and Manuscript Library.

This two-pronged approach represented a compromise for Bourne, between the ideal of true mathematical navigation and the practical realities of practitioners' limited mathematical abilities. The apparent contradiction it contains is resolved in the realization that Bourne sought to address two communities at once, acting as an expert mediator between them. His simultaneous appeal to mathematically minded patrons and unlearned practitioners is a clear example of the expert's unique ability to bridge the gap between two groups with diverse backgrounds and interests, even while reinforcing the issues that divide them. Bourne wanted his reader-practitioners to be able to use the new mathematical technologies, and to understand them if possible, but he did not expect them all to live up to his ideal. By writing *as if for mathematicians*, Bourne encouraged even his most practical readers to begin *thinking like mathematicians*, something desirable if they were to embrace fully the helpful innovations he offered them. Yet by drawing such a distinction between his mathematically capable and incapable readers, Bourne also implicitly redefined navigational expertise as the rightful province of the mathematician, rather than the unlearned pilot, however experienced he may be.

In the end, though, *A Regiment for the Sea* was a manifestly practical book. William Bourne may not have been a mariner himself, but he was still a practical man with whom most mariners would have identified. He had spent his life living in close communion with them and was well aware of their anxieties, needs, and shortcomings: "I know the nature and quality of some that take charge [of ships], they will have instruments and other things thereunto appertaining, and yet they themselves do not know the use of them; yet they will seem to be cunning, and that they need no instructions of any man, for that they know all things, and yet in respect know nothing."[27] Bourne hoped to provide such mariners with a better grasp of the mathematical theories underlying their navigational technologies; but he also realized that in the imperfect world of actual practice, using an instrument successfully and profitably did not always require understanding how it was conceived, designed, and constructed. Using an instrument correctly, even without understanding it, was still better than nothing, and Bourne's strategy of taking the bulk of the manual's mathematical responsibilities upon himself in order to remove the burden from his unlearned readers was probably a major factor in the long-term popularity of *A Regiment for the Sea*.

## *William Borough*, A Discovrs of the Variation of the Cumpas

Ironically, Bourne's emphasis on practical application over theoretical understanding was contradicted most sharply by a practicing navigator of great experience and renown, William Borough, in his treatise *A Discovrs of the Variation of the*

*Cumpas*. As mentioned in chapter 3, Borough began his nautical career as a ship's boy, sailing under Richard Chancellor on England's first voyage to the northeast in 1553, when he was sixteen years old. He gained further experience sailing in Arctic waters under the command of his brother Stephen and went on to lead a number of northeastern trading voyages himself. But William Borough was no common seaman or ship's master; he was also a talented mathematician and was deeply committed to propagating mathematical technologies of navigation. During the 1570s, in between his regular voyages to Russia for the Muscovy Company, he tried to pass on some of his applied mathematical knowledge to Martin Frobisher, Charles Jackman, and Arthur Pett as they prepared for their respective voyages of exploration. Like Bourne, Borough declared that his navigational instruction was intended for the benefit of practicing mariners; but there can be no doubt that he had a more mathematically sophisticated audience in mind. He gave far greater expression to his mathematics than Bourne ever attempted.

Borough's subject matter in *Discovrs* was limited to the magnetic variation of the compass. His treatise was written to complement the work of his friend Robert Norman, another experienced mariner and a maker of nautical instruments. Norman had written his own work on magnetism, called *The Newe Attractiue*..., which he dedicated to Borough. The treatises were published together in 1581, and Borough's went through two subsequent editions during the next fifteen years. As a seaman himself, Borough was certainly well aware of the severe mathematical limitations of the average (and even above-average) maritime reader, but nevertheless he insisted that this was the audience intended for his work, addressing his preface "[t]o the travelers, seamen, and mariners of England." Borough believed that mariners' and cartographers' collective ignorance of magnetic variation was perhaps the greatest single impediment to the making of accurate and useful navigational charts. He was most eager "that no opportunity be omitted, when, or where any observation may be made, either for the variation, or latitude of places, or of any other necessary point incident to navigation." Borough knew better than most that only true seafaring men would ever be in a position to make most of the observations needed so badly, so it was vital that they be taught to do so correctly.[28]

The prevalence of complicated mathematics in his treatise, Borough believed, should not detract from his ultimate purpose in publishing it. His "chiefest intent" in writing the manual was "to pleasure those [mariners] that shall have occasion to put the thing in practice by their own travail and experience," rather than to educate land-bound mathematicians. He acknowledged that his inclusion of so many "rules... deducted from the fountains of the mathematical sciences... which may seem strange in our English tongue" might make it seem that he had "missed of my first good meaning," and he urged his less educated maritime readers to "choose that which is plain, and comfortable to their capacities, and make their profit thereof." At the same time, however, he charged them to look upon the study of mathematics as a cornerstone of their normal training. Borough believed "all seamen and travelers, that desire to be cunning in their profession" should first "seek knowledge in arithmetic and geometry, which are the grounds of all science and certain arts." Such an education had been made readily available, he felt, through the numerous pedagogical treatises of Robert Recorde and others, "written in our English tongue, sufficient for an industrious and willing mind to attain to great perfection." Once properly trained in mathematics, every competent English pilot should be able "not only [to] judge of instruments, rules, and precepts given by other[s], but also be able to correct them, and to devise new of himself. And this not only in navigation, but in all mechanical sciences."[29]

If Borough really meant for his unlearned reader-practitioners to absorb only the portions of his book that were "plain, and comfortable to their capacities," he left them precious little on which to focus. Nearly all of his twelve chapters were heavily mathematical in content and would have required a firm grasp of geometry and trigonometry just to comprehend them. He apparently assumed such a thorough familiarity with Euclidean geometry, for example, on the part of his readers that he not only referred repeatedly to the *Elements* but usually cited definitions and theorems by number only. He also expected a command of both plane and spherical trigonometry, subjects well beyond the competence of virtually all practicing English seamen. At that time there was not even a textbook written in English covering those topics. Borough even referred his readers to the highly technical mathematical and astronomical works of Nicolaus Copernicus, Erasmus Reinhold, Georgius Rheticus, and others. Most of these works existed only in Latin and would have represented a challenge to many university-trained astronomers.

In his sixth chapter, for instance, Borough explained how to find the magnetic variation for a particular place using only a single observation of the sun, made at any time of day, given only the observer's latitude and the solar declination of the date in question:

> For the accomplishing of this proposition, you are to imagine a spherical triangle upon the superficies of the globe, whose sides must be: first the portion or arc of the meridian between your zenith and the pole, which is the complement of the latitude; the second the arc of the vertical circle contained between your zenith and the sun, which is the complement of the sun's elevation at the time of the observation; the

third side is an arc of the circle of declination comprehended between the sun and the elevated pole, this arc is found by adding, or subtracting, the declination of the sun, to or from, the quadrant or 90 d[egrees], which must be done with this consideration, that if you be on the same side of the equinoctial that the sun is, you are to subtract the declination from the quadrant. If on the other side, to add it to the same, so have you the three sides of the spherical triangle given. Then the substance of the work consisteth in finding the quantity of the angle of the same triangle at the zenith, for the complement thereof to the semi-circle or two right angles, is the horizontal distance of the sun's azimuth from the meridian, which being compared with the variation of the sun's shadow upon the instrument, giveth the thing required.

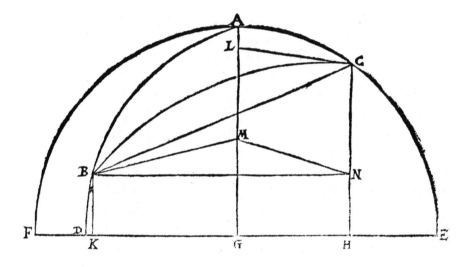

Let FACE be the meridian, wherein A the zenith, C the pole. AD the vertical circle or azimuth of the sun passing by B the place of the sun at the time of the observation. BD the elevation of the sun. BA the complement of the elevation. AC the complement of the latitude. BC the arc of the circle of declination, or the chord of the same arc. FGE the plane of the horizon.

Now from the three angles of the triangle ABC let fall 3 perpendicular lines to the plane of the horizon AG, CH, and BK, and by the 6th of the 11th of *Euclid*, these three lines shall be parallels.

Then let fall a perpendicular line from C upon AG in the point L, from B another perpendicular upon the same line AG at the point M. And from the same point M erect a perpendicular line to N which shall be parallel and equal to LC. Then join B and N together. So have you a right-lined triangle BMN, whose angle at M is equal to the angle A of the spherical triangle ABC. By the 4th definition of the 11th of *Euclid*, for the like reason is of obtuse angles as of acute or sharp. And the sides thereof BM and MN are given, BM the sine of BA and MN equal to LC the sine of CA. And the third side BN is found by subtracting the square of NC from the square of the chord BC, as in the 47th of the first of *Euclid*.

*And in right-lined triangles, the three sides being given, the angles are also given*, by the 44th, 45th, etc. of the first of *Regiomontanus*, and by the 7th proposition of the 13th chapter of *Copernicus* his first book.[30]

In explaining this single exercise, Borough has assumed in his readers a knowledge of spherical trigonometry, a facility with interpreting geometric diagrams, and apparently even ready access to the works of Regiomontanus and Copernicus for further reference—books that probably graced the library of not a single English seagoing vessel in 1581 (except perhaps Borough's). Ironically, it may well have been Borough's extensive experience at sea that led him to introduce so much complex mathematics. After all, he knew as well as any seaman that the weather did not always permit one to make astronomical observations when and where one chose, and so a method that allowed greater temporal flexibility might have been extremely useful to a practicing pilot, at least in theory. Nonetheless, given how far Borough's method went beyond the mathematical competence of virtually every pilot in the English merchant marine, his commitment to helping "the travelers, seamen, and mariners of England" may well be questioned. Rather, like the work of Copernicus to which he referred, *A Discovrs of the Variation of the Cumpas* reads much more like a clever mathematical treatise, written by one mathematician for the gratification of others.

## Thomas Hood

After the death of William Borough's brother Stephen in 1584, the official role of training and educating mariners in and around London shifted from a mathematically minded navigator to a university-educated mathematician, Thomas Hood. Despite Stephen Borough's fervent efforts, there had long been agitation for a more formal, institutionalized lectureship in practical mathematics to be established in London. Borough himself, it will be remembered, had tried and failed to create the office of pilot major of England along such lines in 1562. During the early 1580s, Richard Hakluyt lobbied the Privy Council and other powerful Londoners "again and again" from his embassy in Paris, praising the mathematical lectureship instituted there at the prompting of Peter Ramus. In one letter of 1584,

Hakluyt suggested to Francis Walsingham that two similar posts be created in England—one in Oxford focusing on pure mathematics, and a second in London dedicated solely to practical mathematics: "For it is not unknown unto your wisdom, how necessary for service of wars arithmetic and geometry are, and for our new discoveries and longer voyages by sea the art of navigation is, which is compounded of many parts of the aforesaid [mathematical] sciences."[31] Just as in the 1560s, however, enthusiasm was slow to build, and the necessary funding could not be procured until the threat of the Spanish Armada in the summer of 1588 stimulated a keen interest in the martial applications of mathematics.

The London mathematical lectureship was finally instituted in the autumn of 1588, after the Armada had already been scattered, by a group of prominent London merchants led by Sir Thomas Smith; they chose Thomas Hood to serve as their first lecturer.[32] During the 1580s, Hood had graduated with a master of arts degree from Trinity College, Cambridge, where he then became a fellow, and had briefly served as the university's mathematical lecturer; he also held a medical degree entitling him to practice as a physician anywhere outside of London. He held the London mathematical lectureship for four years, from 1588 until 1592, having been reappointed once in 1590.[33] The lectureship was originally intended to provide mathematical training not for sailors but for the captains of London's "trained bands," the civilian militia that would serve as England's first and only line of land-based defense in the event of an invasion. Hood was supposed to focus accordingly on topics of importance mainly to the infantry and artillerymen, such as surveying, gunnery, and fortification.[34] After 1588, though, as the threat of an immanent Spanish attack faded, navigation soon became Hood's dominant theme.[35]

Although the exact nature of the audience Hood addressed in his lectures remains largely unknown, the style and content of his printed works, most of which were based to some extent on his lectures, provide some clues regarding his most likely auditors. During the 1590s, Hood published a series of vernacular treatises on a wide range of practical mathematical subjects, including the use of the celestial and terrestrial globes (the first such works in English), the making and use of his own modified cross-staff, and the proper use of the plane chart in navigation; many of these went through several editions. He also translated into English a textbook on arithmetic by Christian Urstitius and another on geometry by Peter Ramus and edited the 1592 edition of William Bourne's *A Regiment for the Sea* (Hood's edition was reprinted in 1596 and 1611).[36]

Despite his Cambridge origins, Hood's printed treatises (and the lectures from which they derived) would probably not have appealed to academic mathematicians. In accordance with the terms of his lectureship, his works were deliberately aimed at an audience more interested in practical applications than in mathematical sophistication and more comfortable reading English than Latin. Hood tended to focus much more on explaining how to apply the technologies he discussed than on elucidating the complex mathematical framework behind them; in that sense his approach had much in common with Bourne's, with even less emphasis on mathematics for its own sake. In *The Marriners guide*, for example, a short treatise on the use of the plane chart appended to his 1592 edition of Bourne's *Regiment*, Hood was careful to point out that meridians do not actually run parallel to one another on the surface of a sphere. Indeed, he explained, their portrayal as parallel lines on a plane chart rendered that instrument less suitable for use the farther one sailed from the equator, because it badly distorted the coastlines.[37] But having offered the necessary mathematical caveats, Hood did not hesitate to assume that meridians *were* parallel, in order to teach his readers how to use what was still by far the simplest and most common type of chart projection in 1592. Also reminiscent of Bourne, Hood included a number of examples in his lessons that might be taken as models, and thus he kept to a minimum the degree of mathematical mastery his readers needed to possess. He even engraved his own plane chart of the eastern Atlantic Ocean, intended for publication with *The Marriners guide*, upon which he expected his readers to work through his sample exercises, many of which referred explicitly to courses he had already pricked out on the chart.

Was Hood lecturing and writing for a practical maritime audience, then? This was how he often portrayed himself. He explicitly addressed *The Marriners guide* to "the industrious sailors," for example, and claimed to have a personal knowledge of their particular needs and concerns: "I have had to do a long time with diverse of your profession, both for the making of sea cards, and also for instructing them in mathematical matters belonging to navigation. Amongst whom I have found many willing to learn, and by that means had an insight into their wants." Yet if Hood was not writing for a purely mathematical audience, he was not writing for a thoroughly nautical one, either. His pedagogical style often seems less suited to common soldiers and sailors than to a small but influential circle of London merchants and gentlemanly patrons, with an amateur interest in applied mathematics. In the very same paragraph as the above quotation, for example, Hood referred to his readers as "*Gentlemen* Sailors, and whatsoever else you are that travail by sea."[38] Very few sixteenth-century English seamen could have called themselves "gentlemen," and among Hood's most likely readers, the latter category was probably the more operative one.

The captains of the trained bands in London, for whose benefit the lectures were originally instituted and funded, were usually the city fathers—members of

*Figure 4.3.* Thomas Hood's engraved chart of the eastern Atlantic Ocean, meant to be included with his short treatise, *The Marriners Guide*. Hood expected his readers to use the chart as a platform on which to work through the numerous examples included in his text. The chart itself contains sample courses that are explicitly referred to in the text. The map is inscribed "T. Hood descripsit. ARyther sculpsit 1592," but this particular example is bound with the 1596 edition of *The Marriners Guide*. Photo courtesy of the Peterborough Cathedral Library.

London's gentlemanly and mercantile elite—and such individuals most likely made up the vast majority of Hood's audience. In his dedication of his 1590 translation of Peter Ramus's *The Elementes of Geometrie* to the lord mayor and aldermen of London, Hood claimed that his lectures had benefited "not only those young gentlemen, whom commonly we call the captains of this city . . . but also all other whom it pleased to resort unto." Yet even these "all other" that Hood mentioned were not practical men but "sundry grave and wise men," members of London's ruling classes.[39] Moreover, although most of his lectures were delivered at the Staplers' Chapel in Leadenhall Street, his inaugural lecture was given at the private home of Thomas Smith—a venue that suggests not a large public gathering of soldiers and sailors, but rather a small, select group of wealthy auditors. The mathematical lectures were explicitly intended for the good of the public and were ostensibly open to all who wished to attend them, but the humbler sort of practitioners were probably not well represented among Hood's listeners. Nor would they have found much of use to them in his printed works, the style and tone of which seem more appropriate for those who invested in ships than for those who sailed in them.

Despite his position as a lecturer, Hood's favorite written pedagogical style was the dialogue, usually between a Master ("M.") and a pupil called "Philomathes." In *The Marriners guide,* Philomathes was clearly no apprentice mariner practicing his craft; he was a one-time passenger of some social standing, who wished to learn more about the charts he had witnessed some of the sailors using during a recent journey. He approached the Master already aware of many of the strengths and weaknesses of the plane chart, having discussed its use with the ship's crew, though the seamen were too uneducated to explain everything to him sufficiently. It was perhaps a defensive rhetorical point on the part of Hood, a university mathematician who had never been to sea himself, that Philomathes had failed to gain a satisfactory understanding from experienced practitioners and resorted instead to the land-bound mathematical Master for further enlightenment. For his part, Hood's Master showed some token respect for the mariners' applied knowledge, hoping "that this my deed be not prejudicial to any other man's, whose experience in hydrographical matters is more than mine." Yet at the same time, he asserted the right of the academic mathematician to address practical matters, even without personal experience, implying that an unlearned practitioner was unlikely to do so himself in any rigorous fashion: "[W]hen his discourse cometh forth, let it be accepted as it shall deserve; in the mean season, let this serve the turn."[40]

Hood's use of the dialogue format was most effective in explicating the finer points of complicated subjects to a nonmathematical audience, by allowing the student Philomathes to ask pertinent questions and receive clearer, more detailed instruction from the Master. Certain elements of these fictional lessons, however, suggest that they were less suited to a ship master's cabin than to a private study, where Hood himself was much more at home. In addition to his public lectures, Hood offered personal tutorials to any patrons willing to supplement his stipend. Such arrangements were relatively common during the sixteenth century, as wealthy aristocrats and gentlemen in need of some sort of specialized knowledge offered patronage to those who possessed it. In this case, investors in English mercantile companies and voyages of exploration turned to Hood to give them a better understanding of the mathematical technologies that they believed improved their chances of realizing a profit. It is telling that the setting, format, and style of *The Marriners guide* and Hood's other dialogues resemble closely the sort of private, one-on-one tutorial available only to London's wealthy elite.[41]

Hood also presented his treatises in such a way that conveniently accessing specific lessons was difficult, if not impossible, for the reader-practitioner. For example, his book entitled *The Vse of Both the Globes, Celestiall, and Terrestriall* . . . (1592) was roughly two hundred pages long but did not include such helpful tools as an index, a table of contents, or even folio numbers. In order to refer to some lesson that he might need, the reader would have to leaf through the text until the passage in question was found, or else rely upon his own marginal notes.[42] Although the lack of an index or a table of contents was certainly not unusual in sixteenth-century printed texts, such a treatise would have been most inconvenient for actual use at sea, where the ability to reference specific information handily might well have been vital for an aspiring navigator. In contrast, William Bourne provided a table of contents at the end of his *A Regiment for the Sea* that even included page number references.[43] Bourne also gave detailed headings to each of his chapters, summarizing their contents, whereas Hood's dialogue simply presented an undifferentiated and unbroken string of questions and answers. Finally, a book as large as *The Vse of Both the Globes* would have been very costly and probably well out of the price range of most practicing seamen, and in 1592 few English ships even carried globes as navigational equipment, giving mariners little effective reason to read the book in any case.

Given their emphasis on basic application over thorough comprehension, together with their modest level of mathematical sophistication, Hood's lectures and treatises were probably not meant for an audience of mathematicians. Yet this did not necessarily mean that they were written for the benefit of "the industrious sailors," as Hood claimed. Rather, with their pedagogical style reminiscent of a gentlemanly tutorial, and their veiled disdain for the needs and abilities of practicing seamen, Hood's treatises were probably meant to be read and enjoyed by London's noble, gentlemanly, and mercantile classes. These were, after all, exactly the same wealthy patrons who commissioned and paid for his public and private mathematical instruction in the first place.

## *John Davis*, The Seamans Secrets

The renowned explorer and navigator John Davis combined in one person the nautical experience of a seasoned mariner like William Borough and the rare mathematical talent and clear pedagogical style of Thomas Hood. During his long maritime career, Davis guided English ships all over the world, from his pioneering explorations of the coast of Greenland in the mid-1580s to the maiden voyage of England's East India Company in 1601. Few officers in England, or indeed in all the world, could boast more practical experience in plotting and following a nautical course, in both known and unknown waters. In 1594, one of the rare years in which he was not at sea, Davis composed one of the most practically oriented treatises on the art of mathematical navigation, *The Seamans Secrets*.[44] Like Bourne and Hood, Davis usually stressed practical application over deeper comprehension, but the theoretical underpinning of his lessons was never far from his mind. Although he simplified some of his mathematics for the sake of unlearned reader-practitioners, he was always careful to note when he did so. At the same time, he occasionally introduced more complex mathematics that he acknowledged would be of little use to the average navigator at sea, simply for the edification of the more advanced reader.

Davis's affinity for the more applied branches of mathematics, together with his own considerable experience in working with actual seamen, allowed him to appreciate the limited mathematical ability and patience of his reader-practitioners, as well as their need for the most basic and immediately useful kinds of instruction. He composed his manual accordingly, eschewing formal mathematical prose in favor of a more pragmatic, straightforward style: "To manifest the necessary conclusions of navigation in brief and short terms is my only intent, and therefore I omit to declare the causes of terms and definition of artificial words, as matters superfluous to my purpose, neither have I laid down the cunning conclusions apt for scholars to practice upon the shore, but only those things that are needfully required in a sufficient seaman."[45] Davis presented his lessons in a simple but effective question-and-answer format, reminiscent of Hood's dialogues but without named interlocutors. His book was relatively popular, with four subsequent editions during the first half of the seventeenth century.

As William Bourne had done before him, Davis concentrated on practical application; he usually provided a mathematical explanation for the technologies he presented, but his main concern was that his readers be able to put them to use. Toward that end, he included several tables and instruments in his text—black boxes, to reduce or eliminate the reader's need for tedious calculations by doing them for him beforehand—such as those for finding the lunar epact, the times of high tide, and the distance that must be sailed along any given heading to alter one's latitude by one degree.[46] Moreover, in virtually every chapter of his text, he incorporated literally dozens of examples of varying complexity, far more than any previous author had provided. These examples, which included both the problems and their solutions, could be used as models for the reader to refer to whenever the need arose. For instance, in illustrating how to use a particular textual instrument, designed to determine one's latitude from the sun's altitude and declination, Davis presented this sample calculation (along with three others):

> 4Q. The sun having 12 degrees of south declination, and being upon the meridian south from me [i.e., at local noon], is 30 degrees above the horizon, I demand how far the sun is from my zenith, how much the equinoctial [equator] is above the horizon, and what is the pole's height [i.e., the observer's latitude].
>
> 4A. First I bring the thread to the place of the sun's declination as before, there holding it not to be moved, then I turn the horizon until I bring it to be 30 deg. under the thread, and then the thread showeth me that the sun is 60 deg. from my zenith, and the horizon showeth that the equinoctial is 42 deg. above the same, and that the north pole is also elevated 48 deg. above the horizon.

Although Davis acknowledged that such examples were often "very easy and plain" and might "readily be answered by memory," he thought it worthwhile to include them nevertheless, "for the benefit of such as are not altogether expert in these practices," so that even his less advanced readers might "gather thereby the more sufficient judgment in this part of navigation."[47]

Beyond providing copious examples, Davis did everything he could to make the mathematics in his text accessible; he explicitly simplified or omitted certain material for the ease of his unlearned readers. When discussing the phases of the moon, for example, Davis explained that a complete lunar cycle lasted 29 days, 12 hours, and 44 minutes, but he also wrote that most mariners commonly assumed that the moon lost 48 minutes per day. This worked out to a cycle of 30 days, which was 11 hours and 16 minutes too long, but he considered this discrepancy too minor for the average mariner to worry about it: "[B]ecause this difference breedeth but small error in their account of tides, therefore to alter practiced rules where there is no urgent cause, were a matter frivolous, which considered, I think it not amiss that we proceed therein by the same method that commonly is exercised." Davis also provided two different methods for calculating the time of high tide, one of which required no multiplication, for use by "those that are not practiced in arithmetic." He described a trigonometric means for calculating the sun's declination but did not recommend its use because of the advanced mathematics involved: "because seamen are not acquainted with such calculations, I therefore omit to speak further thereof, since this plain way before taught is sufficient for their purpose." In short, for his part Davis conceived of navigation almost entirely in mathematical terms, even to the point of hinting at his creation of a means of "navigation arithmetical" that relied entirely upon plane and spherical trigonometry in place of charts and globes.[48] Yet at the same time, he did whatever he could to make the practical mathematics of navigation as palatable and manageable as possible for common seamen.

## *Edward Wright*, Certaine Errors in Navigation

Very little was wrong with the art of navigation, according to the Cambridge mathematician Edward Wright, that could not be fixed by a more thorough and precise application of mathematics. Wright was a fellow of Caius College and a skilled mathematician whose later works covered a host of mathematical arts, though he concentrated especially on navigation. He eventually succeeded Thomas Hood as London's mathematical lecturer when that post was restored in the seventeenth century, taught mathematical navigation to the pilots of the English East India Company, and even tutored the young Prince Henry, son of King James I. Yet Wright was still just an obscure Cambridge fellow in 1589 when George Clifford, Earl of Cumberland, called him to service at sea on an expedition to attack and plunder Spanish ports in the Azores, just one year after the defeat of the Armada. While serving on the mission as a navigator—apparently his first and only experience at sea—Wright became aware of numerous shortcomings in navigational practice which he determined to correct mathematically.[49] The resulting work, entitled *Certaine Errors in Navigation*, was mostly completed by 1592, though Wright did not publish it until 1599, by which time portions of the work had already been printed, both with and without his permission, and were sometimes being claimed by rival authors.[50]

In his preface to the reader, Wright showed himself to be sensitive to criticism regarding his relative lack of nautical experience; he countered by going on the offensive. Those who believed that it was "beyond a land man's skill, to find faults in matters belonging to the sea man's art and profession" had the matter backward, he argued. A firm grasp of mathematics was worth at least as much in mastering the art of navigation as a lifetime's experience at sea. Instruments such as the plane chart, the cross-staff, and the astrolabe, standard equipment for virtually every English pilot and ship's master by 1599, were fundamentally *mathematical* devices, after all. As such, Wright declared, they might "by him that hath been but meanly conversant in mathematical meditations be better apprehended, than otherwise [they] can by the seafaring man, though he spend his whole life in sailing over all the seas in the world."[51] In other words, the art of navigation was not only an appropriate arena for the mathematician to exercise his skills; it was truly a mathematical discipline to begin with, one that practicing seamen grasped but dimly in their mathematically ignorant wanderings.

*Certaine Errors* was unapologetically a work of mathematics first and foremost; in some ways its navigational focus might even be viewed as an incidental, rather than an essential, feature. Regarding the sophisticated mathematics and astronomy that "might have been omitted, as being impertinent to the use of mariners, and exceeding their capacity," Wright argued that it was simply not possible to discuss navigation *properly* without far outstripping most seamen's poor mathematical abilities. Nor was Wright interested in writing primarily for mariners, whose mathematical ineptitude appalled him; his was a book written explicitly for the enjoyment of other mathematicians like himself. "[I]t was not my purpose," he wrote, "neither could I in all places, apply myself to the most part of seamen's capacity: knowing many that would not be content with this regiment alone, but that desired more to know the root from whence this fruit grew: whose desire I was also willing to satisfy as I could for the present."[52] Mathematics, as Copernicus had asserted in 1543, was for the mathematicians and should be left to their judgment and appreciation; the art of navigation, as an applied branch thereof, was to be no exception.

Wright's desire to appeal more to mathematicians than to maritime practitioners is evident in his tendency to use mathematics to make navigation even more complicated than it already was. In his criticism of contemporary practice, he introduced a number of alternative calculation methods for various functions, such as finding the magnetic variation of the compass using only one observation of the sun's bearing and altitude, or calculating one's most direct course without a chart, using only latitude and longitude data.[53] These substitute methods were intended to eliminate the need for superfluous observations, or to replace tools such as the chart altogether, by employing still more complex mathematics instead. They were not meant to *simplify* navigation for the unlearned, but to *mathematize* it even further for the adept. In these sections, Wright wrote not for sailors but for the readers he mentioned in his preface who had the rare mathematical knowledge and skill required to follow him and who could therefore best appreciate his talent and ingenuity.

Nevertheless, Wright was careful to organize his manual so that even unlearned seamen might still profit from it. The treatise as a whole addressed four main categories of "error" in late-sixteenth-century navigation, including the geometric distortion of coastlines inherent in the traditional plane chart, due to its use of parallel meridians; the general failure of navigators to account for the magnetic variation of the compass; the many difficulties in trying to use a cross-staff accurately; and the myriad troublesome errors in contemporary printed astronomical tables, which limited the accuracy of all calculations based upon them. In presenting his solutions to these errors, Wright adopted the dual strategy employed by many of his predecessors: he provided detailed explanations of the complex mathematics behind them, yet he also black-boxed many of them, using instruments and tables to remove the more troublesome calculations from the reader's attention and responsibility.

The best solution for almost every kind of navigational or observational difficulty, according to Wright, was a table. He provided the reader with tables of solar declination and right ascension (for calculating local time), as well as the declinations of numerous fixed stars; tables of correction for nearly every common source of error in using the cross-staff, including some that Wright urged mariners to ignore, since they were so subtle as to be irrelevant given the unavoidable imprecision of measurements taken at sea; and tables of magnetic variation, measured by Wright himself during his voyage to the Azores in 1589.[54] By relying upon tables, Wright was able to present his solutions as black boxes, separating their practical application from the sophisticated mathematics that stood behind them. Unlearned mariners could thus use his book to correct their errors in practice, profiting from his new methods without ever grasping how or why they worked. When measuring a celestial altitude with a cross-staff while aboard ship, for example, a pilot need not understand the astronomy and geometry involved in order to use Wright's table and simple instructions to correct for the height of his eye above the surface of the water, a subtle error that could nevertheless result in several miles' discrepancy.[55] Yet Wright never left his tables to stand on their own; he was scrupulous about explaining why and how he had created them, even when his explanations were hopelessly beyond the reach of maritime reader-practitioners, so that his true mathematical peers could admire his every ingenious innovation.

This dual approach of writing for mathematicians and also black-boxing the mathematics for mariners was nowhere more evident than in Wright's discussion of what is now known as the Mercator projection chart. Wright admitted taking inspiration for his own chart from Mercator's famous 1569 world map, though he claimed to have arrived at the mathematical solution for creating it independently. The great innovation of the Mercator projection was that although meridian lines were still portrayed as parallel and equidistant, as on any common plane chart, the lines of latitude were spaced proportionately farther apart from one another; the higher the latitude, the farther apart the latitude lines. This maintained the proper ratio between the *linear* lengths of one degree of latitude and longitude, even relatively far away from the equator. Coastlines at higher latitudes were still grossly exaggerated in size, but they were depicted in proper relation and proportion to one another; this made accurate distance calculation possible, provided that the latitude scale nearest the position in question was used in the process. Furthermore, whereas rhumb lines on a spherical surface spiral toward the poles, on a Mercator projection chart they are accurately depicted as straight lines. This crucial difference made determining the constant compass heading to be sailed between two points as simple as finding the bearing of a straight line on the chart.[56]

The first part of *Certaine Errors* explained how to construct such a chart trigonometrically—the distance between lines at each degree of latitude becomes a function of their secant. Ideally, using Wright's text as a template, the mathematically minded reader should have been able to construct his own chart of any part of the Earth he desired simply by spacing the latitude lines in their proper trigonometric proportion and then placing geographic features based on their correct latitudes and longitudes.[57] This was a feat well beyond the mathematical capabilities of most English mariners of Wright's day. Yet once such a chart had been produced by someone else, the entire mathematical framework underlying it could be ignored by unlearned readers; indeed, the completed chart was even simpler to use in plotting courses than a normal plane chart. There was no reason why a ship's master could not profitably use a trigonometrically constructed Mercator projection chart, even if he had never heard of secants, provided someone else made the chart for him. Conveniently enough, this is exactly what Edward Wright did, including in the manual his own Mercator projection chart of the eastern Atlantic Ocean.

The *inaccuracy* embedded in traditional plane charts had thus been corrected through the increased mathematical *complexity* requisite in making Mercator projection charts. Yet the maritime user had no need to trouble himself with the mathematics behind the new projection, because Wright had already completed the mysterious calculations for him. The same strategy lay behind Wright's multitude of tables; thanks to his great talent and effort in compiling them, even an ignorant pilot could apply sophisticated mathematics to correct his many errors, practicing ever more accurate navigation without understanding why or how. While *Certaine Errors* may have been written to impress mathematicians, then, it was simultaneously structured for the profit of a comparatively unlearned audience.

## *Thomas Harriot, "Instructions for Raleigh's voyage to Guyana"*

At the other end of the spectrum from Edward Wright's *Certaine Errors*, in both style and intended audience, were the navigational instructions compiled by Thomas Harriot in 1595. Harriot was one of the most renowned mathematicians and astronomers of his day, a friend and colleague of John Dee and a correspondent of Johannes Kepler. Like Wright, he had both academic credentials and maritime experience and considered mathematics to be the foundation of the navigator's art. He had received his master of arts degree from St. Mary Hall, Oxford, in 1579, and immediately entered the service of Walter Raleigh as a tutor of mathematical navigation technologies. He sailed with Raleigh as a navigator during the latter's 1585 voyage to explore and colonize the region of North America that is now Virginia and played an integral role throughout that venture. Not only did he help to navigate the colonists' ships across the Atlantic; he also surveyed and mapped the area around the fledgling colony and learned the language of the Native Americans they encountered there well enough to open a communication with them. His short treatise *A briefe and true report of the new found land of Virginia* . . . is one of the earliest and best-known first-person accounts of English contact with the New World.[58]

As with Wright, Harriot's time at sea had convinced him that contemporary navigational practices were rife with flaws and that these could only be corrected through a more precise and mathematical approach to the art. Unlike Wright, however, Harriot composed his instructions not for the casual perusal of other mathematicians but for the practical benefit of his patron, Raleigh, and the other pilots in his employ. During the mid-1580s, Harriot wrote a manuscript manual of navigation that he called "Arcticon," most likely for Raleigh's use in preparing for his 1585 voyage to Virginia. The manual itself was never published but was presented to Raleigh directly, and unfortunately it is now lost. However, as Raleigh readied himself and his fellow officers for a subsequent voyage to Guyana in 1595, he asked Harriot to refresh their memories with respect to the mathematical technologies he had covered in "Arcticon." Harriot obliged them with a series of six informal lectures, the manuscript notes for which survive, together with a few tables compiled for use both with the lectures and at sea. These surviving notes and tables probably reflect the same concepts and instruments that Harriot discussed more fully in his completed manual.

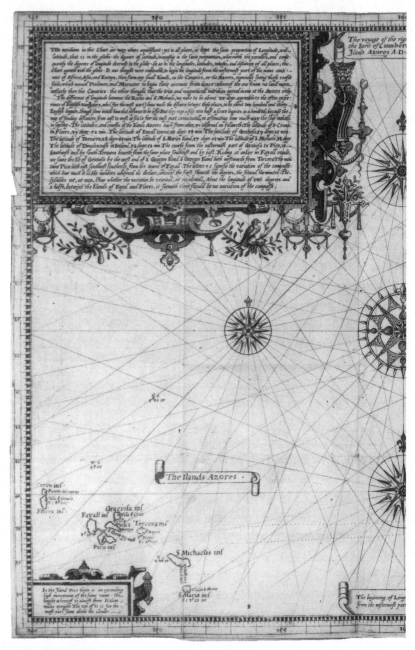

*Figure 4.4.* Edward Wright's Mercator projection chart of the eastern Atlantic Ocean. As with William Bourne's textual instrument (depicted above), the reader of Wright's text did not need to master his explanation of the trigonometric process for constructing such a chart in order to use one successfully, once it had been constructed for him. The

numbers in the upper right portion of the chart are soundings for the English Channel and the western coast of France, a particularly treacherous area in which to sail. Edward Wright, *Certaine Errors in Navigation* . . . (London, 1599). Photo used by permission of the British Library, shelfmark G.7312.

Even in the form of rough lecture notes, Harriot's pedagogical style was truly a model of pragmatic clarity and conciseness. Whereas the complicated explanations in Wright's book were aimed primarily at other mathematicians, Harriot spoke directly to a small, select group of practicing ships' masters who knew for certain they would need the information he was teaching them in the very near future. Raleigh, in fact, had acted as Harriot's patron for so many years in part because he wanted to ensure his own continued access to such useful mathematical instruction at all times. Whereas Wright's intended audience led him to explain in detail the often difficult mathematics behind his many corrective technologies, Harriot most often chose to omit explanations that he considered superfluous to the practical matter at hand. He knew his audience well and adhered for the most part to exactly the kind of basic, functional information that would be of greatest interest and use to them, presenting it in as clear and logical a fashion as could ever be hoped for.

Like many of his fellow authors and instructors, Harriot took responsibility for handling all the thorniest mathematics himself, presenting only the black-boxed technologies to his audience so that they would be little troubled by tedious or difficult calculations. His second lecture, on the errors commonly made in measuring the altitude of celestial bodies with a cross-staff, is most illustrative in this regard. The first error, which Harriot called "parallaxis of the staff," derived from the fact that the end of the staff could not be placed directly on the center of the eye in making the measurement (Edward Wright termed the same phenomenon "parallax of the eye"). He addressed this problem by determining through experimentation that, so long as the observer always placed the instrument on the corner of his eye (as opposed to his cheekbone or the bridge of his nose), the error was consistent, even for different individual observers; he called this discrepancy the staff's "eccentricity." He then provided a solution to the problem by inscribing the magnitude of the eccentricity directly on the staff itself. The user then had only to adjust his own measurement by subtracting the eccentricity from the celestial object's apparent altitude—no further calculation or comprehension was required, only strict adherence to Harriot's simple method.[59]

A second cross-staff error required the observer to correct for the height of the his eye above the surface of the water when making his observation. Like Wright, Harriot addressed this error by means of a table in which he provided the amount to be subtracted for different heights above the water's surface, as well as an estimate of how high above the water various parts of the average ship might be. Once again, no calculation was required on the part of the user, beyond subtracting the quantity Harriot had listed in his table.⁶⁰ Of two other sources of error, the refraction of the air and the "parallaxis of [the sun's] altitude" (solar parallax), Harriot simply dismissed them as too small to be significant: "[B]ecause in your voyage they amount not either one or both to 3 minutes when most, I leave them for another place and time to be uttered."⁶¹ In effect, anyone using the cross-staff according to Harriot's prescribed method would eliminate potential sources of error and gain greater precision, with only a minimal increase in effort or complexity for himself. Harriot had already tackled all of the harder calculations and thus removed them from the reader's attention.

Harriot's approach to correcting the use of the cross-staff obviously has much in common with Wright's. Both men identified the same persistent errors in need of correction, both provided mathematically based solutions, and both used tables as a means of minimizing the reader-practitioner's mathematical responsibilities. Yet the two authors differed markedly in the style and tone of their instructions, based upon the different needs and expectations of the respective audience each one hoped to reach. In correcting for what he called "parallax of the eye," for example, Wright recommended adding a second cross vane to the staff, twice as long as the first, by which a second observation could be made simultaneously to confirm the first.⁶² This more mathematically elegant approach necessitated the addition of a second complete measurement scale to the instrument to accommodate the second cross vane and doubled the work required of the observer. It was a far more troublesome solution for the practical maritime reader than simply subtracting a corrective constant, inscribed right on the instrument itself, as Harriot suggested in his instructions. Likewise, both authors declared that the effects of solar parallax were too negligible to require a correction for practical use, but Wright nevertheless included a corrective table in his manual, insisting that "the rules and grounds of art . . . so much as is possible ought to be without all error."⁶³ This may be seen as another instance of Wright's more purely mathematical focus, compared with Harriot's more pragmatic point of view.

Perhaps Harriot's greatest advantage over other navigational instructors was his lucid organization and presentation of inherently complex information, as in his fourth lecture on the calculation of one's latitude using the sun's altitude and declination. Explaining this basic yet tricky procedure clearly and concisely had proved more than a match for Martín Cortés, and William Bourne had only improved upon Cortés's attempt by quadrupling the number of chapters and pages he devoted to it. In his own treatment of the subject, Harriot managed to avoid falling into the same quagmire of confusion by taking a new approach. After briefly outlining the main points of the technique, he rejected simple prose as an inadequate means of teaching it, criticizing previous authors such as Cortés whose explanations had been too short, vague, or corrupt to be of much use:

> But whereas [some authors] endeavor thereby to give general rules, besides the obscurity they contain with their brevity, they are in some cases false and insufficient.
>
> I will therefore in another manner set down those rules, not with too naked and general terms as others have done, whereby cannot be yielded the thing desired, without special pausing and troublesome consideration: but in their own proper terms, although they be a few more, yet for that cause more easily and speedily to be understood. And the order of them according to the several cases that may happen, *I think most ready for the pen to express in a table* as followeth.[64]

Rather than attacking the problem in prose alone, then, Harriot chose instead to compile another of his illustrative tables, in which he organized the various choices to be made by the reader into a sort of flowchart, probably inspired by the pedagogical tables of Peter Ramus. By presenting a short series of simple choices, Harriot's table led the reader directly to one of several endpoints that told him exactly which arithmetical calculations he needed to use.[65]

Finally, Harriot included in his notes several examples that reinforced the techniques he was teaching and also served as models by which his auditors could guide their own calculations, should they become confused when applying them later. Harriot's use of examples was somewhat different from that of other authors, in that he usually made them cumulative. That is to say, in an example illustrating a lesson from one of his later lectures, he made sure to include material, calculations, and the use of tables and instruments from his previous lectures as well. He thereby reinforced for his pupils the entirety of his instruction to that point, rather than just the lesson at hand, and forced them to work through calculations more representative of what they would actually have to make at sea. For instance, in the first of two examples that Harriot included with his fourth lecture, on calculating one's latitude using the sun's altitude and declination at local noon, the reader was forced to recall material from each of his previous three lectures:

> A precedent of the first rule, the declination of the sun being north and zenith to the northward of the sun.
>
> Anno 1595, March 18, to the westward of England 900 leagues or 3 hours.
>
> ---
>
> Apparent meridian altitude of the higher edge of the sun by the shorter staff: 79° 35′
> Parallax of the staff    1° 35′}

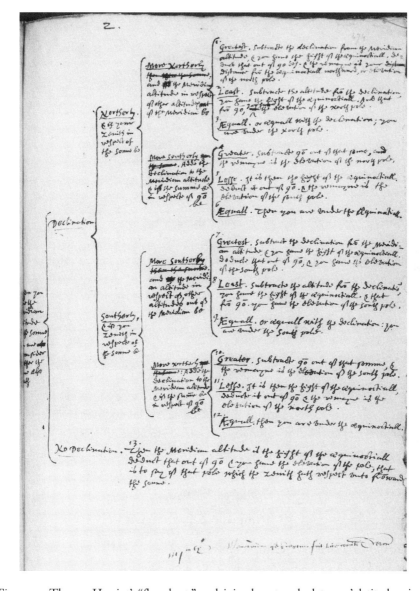

*Figure 4.5.* Thomas Harriot's "flowchart," explaining how to calculate one's latitude using the sun's altitude and declination. The calculation itself was not difficult; the challenge was to know which operations to perform in which cases. Explaining the process in prose, as several previous authors had attempted, often made the question exceedingly confusing. Harriot's flowchart, however, allowed the reader to make his decisions one at a time, using information that he either already possessed or could easily obtain, leading him in the end to one of several brief, simple instructions. Photo used by permission of the British Library, Additional MSS, 6788, fol. 474 r.

| | | |
|---|---|---|
| Surplus of the horizon | 0° 5' } to be abated therefore | 1° 56' |
| Semi-diameter of the sun | 0° 16'} | |

[material from the second lecture, on the use of the cross-staff]
Therefore the true meridian altitude of the sun as by the Astrolabe or ring    77° 39'

[material from the first lecture, on the use of the astrolabe]

───

| | |
|---|---|
| The declination of the sun in the regiment | 2° 56' |
| The part proportional to be added, for 900 leagues | 0° 3' |
| The true declination northerly | 2° 59' |

[material from the third lecture, on finding the sun's declination in a table]

───

And because the zenith is more northerly than the sun, by the first rule abate it out of the true meridian altitude

| | |
|---|---|
| Therefore the altitude of the equinoctial | 74° 40' |
| Deduct out of 90° | |
| Then the altitude of the north pole is | 15° 20' |

[material from the fourth lecture, on finding one's latitude using solar altitude and declination]⁶⁶

Harriot's comprehensive approach made it easier for his auditors to integrate all of his individual lessons into a single, practical method of navigation—a method for which mathematics was the foundation, even though the most complicated aspects of it were carefully removed from the auditor's attention through tables and other well-established black-boxing techniques.

### *Thomas Blundeville*, M. Blvndevile, His Exercises . . .

The ultimate sixteenth-century expression of the newly mathematized art of navigation came not from a mathematically minded mariner, nor from a university mathematician, but rather from a writer and compiler of basic vernacular textbooks, Thomas Blundeville. Blundeville was born into a Norfolk gentry family, a polymath who migrated to London to serve as a tutor in the households of wealthy patrons. He may have been educated at Cambridge, but evidence supporting this is scant. He wrote and translated a number of pleasant, learned treatises on a wide variety of subjects, including philosophy, logic, history, and mathematics, but was probably best known during his lifetime for his original works on horsemanship. In 1587 Blundeville was asked by a young noblewoman he was tutoring, Elizabeth

Bacon, to compose a simple textbook, "so plain and easy as was possible," to help her learn arithmetic.⁶⁷ Although his pupil's chronic illness did not permit her to profit from the resulting work, others who saw it requested copies for themselves, which eventually prompted Blundeville to publish it in 1594, under the title *M. Blvndevile, His Exercises*. . . . In the meantime, however, he decided to expand his modest original text into a lengthy pedagogical tome. Adding considerably to his short work on arithmetic, Blundeville composed and compiled a series of treatises on all manner of mathematical subjects of contemporary interest. The final work was much more comprehensive and was aimed at "those that are desirous to study any part of cosmography, astronomy, or geography, and especially the art of navigation, in which without arithmetic . . . they shall hardly profit."⁶⁸

Blundeville wrote his manual neither for sailors nor for mathematicians, but rather for the same courtly readers who enjoyed his books on philosophy, history, morality, and horsemanship. He dedicated his *Exercises* "to all the young gentlemen of this realm," stating on his title page that it was "very necessary to be read and learned of all young gentlemen that have not been exercised in such disciplines," and addressed the book to "our English gentlemen" consistently throughout.⁶⁹ On the one hand, mathematically advanced readers would have found little of interest in the book, as its contents were for the most part quite basic and derivative from other works. Practicing seamen, on the other hand, would probably have lacked the financial resources needed to purchase such a large and costly textbook, let alone the myriad expensive instruments that Blundeville discussed and recommended. Nevertheless, despite its narrowly focused readership, the book was very popular; a second edition was printed after only three years, in 1597, with six subsequent editions by 1638.

Blundeville's first edition of *Exercises* was a compilation of six individual treatises, each dedicated to a separate topic: the original textbook on arithmetic; an introduction to the principles of cosmography; a treatise on the use of the celestial and terrestrial globes, specifically dealing with the globes of Mercator and Emery Molyneux; a detailed description of the 1592 world map of Peter Plancius; a description and brief set of instructions for using John Blagrave's "mathematical jewel"; and finally a treatise on "the first and chiefest principles of navigation."⁷⁰ Blundeville also appended to his first arithmetical treatise a series of trigonometric tables (of sine, tangent, and secant), together with instructions for their use. In its original form, the book was more than 350 folios long. Later editions included two additional treatises—a reprint of Blundeville's earlier work on the use of maps and charts and an explanation of the tables of latitude and longitude found in Ptolemy's *Cosmographia*—and were thus even longer. Such a massive volume was unmistakably a luxury item, far too expensive for the average mariner to purchase and not very convenient for transportation and use at sea. The very size of his book is a strong indication that Blundeville meant for it to be used in the gentleman's study, rather than in the ship master's cabin.

The treatises themselves were usually little more than a compendium and summary of previously written works, to which Blundeville added a few of his own innovations; this was especially true of the treatise on navigation.[71] He borrowed liberally, though with acknowledgment, from Cortés, Bourne, Norman, Borough, Hood, Wright, and others. Much of the text was simply paraphrased from the original sources, to which Blundeville frequently directed his readers for more comprehensive treatments. In his discussion of finding one's latitude using the sun's meridian altitude and declination, for example, Blundeville advised his readers to consult chapters 7–10 of Bourne's *A Regiment for the Sea* and went on to paraphrase Robert Norman's short treatment of the subject in *The Newe Attractiue*. . . .[72] What set Blundeville's work apart from its predecessors was not its content but its overall organization—indeed, the very conception of the collection itself. Many previous authors had argued that the art of navigation could be neither properly understood nor profitably used without a knowledge of mathematics; yet the rhetorical impact of Blundeville's great tome was both visually and intellectually compelling. Here, for the first time, an elementary treatise on arithmetic was linked through a logical, even inexorable progression to a work on practical navigation. The boundary between pure and applied mathematics was stretched and blurred so that navigation had become just another branch of the greater mathematical tree. The treatises also built upon one another, so that studying the treatise on navigation would have been confusing and frustrating without first mastering the previous five. In fact, in his preface Blundeville urged his readers to tackle the treatises "in such order as they are before set down," or at the very least "to begin with my arithmetic."[73]

For Blundeville, as for many of his fellow authors and their students, navigation *was* mathematics, just as much as geometry and astronomy were. The rhetorical effect of his impressive volume in connecting the various mathematical branches together was all the more striking considering that Blundeville, like several of his colleagues, had never even been to sea; nor indeed had most of the gentlemanly readers for whom he wrote. His was the work of a land-bound, court-centered, vernacular mathematician, who asserted his expertise and authority in what had only recently been an empirically based craft practice, taught through years of apprenticeship. By the time Blundeville published his book in 1594, the art of navigation had been reduced (or, as many authors might have argued, *restored*) to its mathematical foundations. As such, it had become an appropriate subject for educated gentlemen (and, remembering Elizabeth Bacon, perhaps ladies as well) to pursue as amateurs, with the help of experts like Thomas Blundeville.

## Cui bono? The Spoils of Mathematical Expert Mediation

As mentioned above, navigation was but one of the many arts in which English mathematicians asserted their expertise and authority toward the end of the sixteenth century. Ultimately they assumed the role of expert mediators and came to be regarded as valuable servants of queen and commonwealth. How did they succeed in acquiring such a lofty and powerful position for themselves? They had certainly not always been so highly regarded. According to Stephen Johnston, even educated Englishmen during much of the sixteenth century generally viewed mathematics as "a dry and difficult pursuit," its students "a vain or solitary breed, with little sense of civic duty." And this was not the worst that might be said of the subject; beyond being seen as a potentially antisocial waste of time, mathematical study was also associated by many Elizabethans with the Catholic, scholastic tradition at medieval Oxford, with Jesuit scholars and missionaries on the Continent, and with occult pursuits deriving from the Hebraic Cabala. Mathematics was therefore perceived in some circles of post-Reformation England to be connected with popish, heretical, nefarious, and possibly even criminal activities.[74]

Some English proponents of the mathematical arts complained in print about their contemporaries' low opinion of their endeavors. The London-based surveyor Edward Worsop, for instance, in his 1582 treatise on the errors commonly made by surveyors ignorant of mathematics, vehemently denied any connection between mathematics and antisocial behavior. "Most men wrongfully conceive," he wrote, "that certain unlawful practices attributed to astrology are parts of the mathematical sciences, which chiefly bringeth such great discredit, and contempt of mathematicians, and of the pure, and single mathematicals." This led most Englishmen, he lamented, to regard earnest students of mathematics as "fantastical, and vain fellows" at best, "felons and reprobates" at worst.[75] The challenge, for Worsop and others, was to find a way to rehabilitate mathematics, to elevate the discipline in public esteem and improve the reputation of its devotees.

The key to promoting a more positive lay opinion of mathematics, many believed, was to portray it as being not only beneficial and profitable for its students but vital to the well-being of the commonwealth as a whole. Only this could justify the enormous investment of time and effort that mastery of it required and cleanse it of its heretical and occult associations. Worsop, for example, defended mathematics to its critics by stressing the many worthwhile practical ends to be obtained from the study thereof. He pointed out that while his countrymen were all too eager to praise the marvelous achievements of foreign craftsmen, they remained willfully ignorant of the mathematical technologies that were at the root of their success. He boldly connected the profits of mathematical study with scholarly wisdom, noble virtues, and even divine revelation:

> We greatly esteem artificial strangers, for their devises, and workmanships: but we respect not the causes why their doings be more excellent than ours. . . . Great reasons, proofs, and authorities, as well from divinity, as from philosophy, and natural reason, might be brought [to attest] how worthy and needful the mathematical knowledges are. . . . The greatness of the profit, and the judgments of the wisest princes, and of the best learned . . . ought to move the ignorant to detest the hearing, and using, of such doltish and scoffing reproofs, as are used against good arts, so excellently set out, as it were by divine inspiration.[76]

Such interest in the utility of mathematical learning was not limited to tradesmen like Worsop; some of the most advanced educational philosophers of the day made a similar argument. Although English humanists were often reluctant to advocate pure mathematical studies wholeheartedly, they did endorse a moderate mathematical curriculum, so long as it was geared toward practical application and public benefit. In his 1570 treatise entitled *The Scholemaster*, for example, Roger Ascham, one of Tudor England's most prominent humanist educators and a tutor to Queen Elizabeth in her youth, cautioned his pupils against studying mathematics too strenuously for its own sake. Ascham himself was certainly no stranger to mathematical pursuits; while at Cambridge, he had tried (unsuccessfully) to obtain a lectureship in the subject. He insisted, however, that mathematics should always be undertaken as a means to some practical, public end, rather than as a privately pursued end in itself—otherwise, it might prove detrimental to the student's mental and social well-being: "Some wits, moderate enough by nature, be many times marred by over-much study and use of some sciences, namely, music, arithmetic, and geometry. These sciences, as they sharpen men's wits over-much, so they change men's manners over-sore, if they be not moderately mingled, *and wisely applied to some good use of life*. Mark all mathematical heads, which be only and wholly bent to those sciences, how solitary they be themselves, how unfit to live with others, and *how unapt to serve in the world*."[77]

Other educational theorists agreed. Humphrey Gilbert, the Elizabethan courtier, soldier, and mariner whose outline for a new English academy is discussed in the introduction to this book, did not neglect mathematical studies in his idealized curriculum. The students at his proposed academy, Gilbert believed, should be expected to learn some arithmetic, geometry, astronomy, and cosmography. Toward that end, the faculty were to include not just one, but two readers in mathematics. As with the rest of his curriculum, however, practical application was the ultimate goal; Gilbert endorsed mathematical studies not for their own sake, but as necessary preparation for "matters of action meet for present practice, both of peace and war." This pragmatic focus was evident in Gilbert's description of the two mathematical readerships, one of which was to be dedicated to "embattlings, fortifications, and matters of war, with the practice of artillery," and the other to "the art of navigation, with the knowledge of necessary stars, making use of instruments appertaining to the same."[78]

Elizabethan mathematicians, then, had everything to gain by publicly promoting the utility of their subject, making it appear as useful and valuable as possible, even to the point of indispensability. They accomplished this by taking up contemporary technical problems, reinterpreting them in mathematical terms, and then applying their own mathematical skills to develop new solutions for them. Whenever they found themselves in competition with traditional craft methods espoused by common practitioners, they asserted the superiority of their own mathematical solutions. Traditional methods, many argued, were both limited and limiting; borne of concrete experience and fundamentally local in nature, the craft approach only worked for well-known problems and could not be easily adapted to new or unfamiliar circumstances. Mathematical technologies, in contrast, because they were based upon general, abstract principles rather than specific, local circumstances, were more widely applicable and so more valuable to whoever controlled them. By giving its adepts a more fundamental understanding and thus a greater portability of knowledge, mathematics became the only acceptable (and often the only conceivable) means of interpreting a whole host of problems.

In seizing upon such timely issues and offering useful technologies that they alone could provide, English mathematicians represented themselves as a new, essential class of expert mediators, an obligatory point of passage between practitioners and patrons; without help from experts, a complete understanding of the arts in question was impossible, they claimed. Yet their mediation ultimately proved to be divisive; as mathematicians aggressively asserted their control over navigation, for example, they made the art itself much more complex and actually wedged mariners and merchant-investors further apart by inserting between them an indispensable layer of mathematical expertise. Patrons and practitioners could not easily communicate across this new intellectual divide, based as it was upon mathematical knowledge that few on either side possessed without recourse to the mathematicians who had created it. In the process, the mathematicians convinced the patrons whom they cultivated that they were rendering a vital economic and military service to the commonwealth.

Practicing pilots and ships' masters, in contrast, were aware that they had much to lose in conceding navigational expertise to the mathematicians. In doing so, they faced an unpalatable choice: either they could accede to the mathematicians' demands and become minor mathematicians themselves, in order to be considered truly proficient in their own art; or else they would be reduced to the humble status of mechanicians, carrying out by rote the simplified, black-boxed instructions of their mathematical betters. Even if they learned to apply the new technologies, most practitioners would still be forced to settle for a diminished social and intellectual status, unless they also understood the mathematical principles by which those technologies were generated. After all, simply using a mariner's astrolabe or a table of solar declination was not at all the same as conceiving and creating it, or even truly comprehending how it worked. This distinction placed a low ceiling on the degree of expertise that mariners could claim as mere end-users of the technology in question.

Not all practitioners of the newly mathematized arts quietly accepted the reduced status left to them by the mathematicians. Some of the more practically minded authors of treatises mounted a spirited defense of the important knowledge and skill that could still be obtained only through hands-on experience. Robert Norman, for example, was a mathematically gifted but nevertheless undereducated mariner and instrument maker; he did not see himself primarily as a mathematician, and he resented the high-handed way that more formally educated authors dismissed the abilities of craft practitioners. In his preface to *The Newe Attractiue* . . . , he wrote an impassioned argument defending the ability of mariners and artisans to compete with the mathematicians on their own terms:

> And albeit it may be said by the learned in the mathematicals, as hath been already written by some, that this is no question or matter for a mechanician or mariner to meddle with . . . for that it must be handled exquisitely by geometrical demonstration, and arithmetical calculation, in which arts they would have all mechanicians and seamen to be ignorant, or at least insufficiently furnished to perform such a matter. . . . [Y]et there are in this land diverse mechanicians, that in their several faculties and professions, have the use of those arts at their fingers' ends, and can apply them to their several purposes, as effectually and more readily, than those that would most condemn them.[79]

Likewise, in 1598 William Borough wrote disdainfully of his land-bound mathematical rivals, claiming that "none of the best learned in those sciences mathematical, without convenient practice at the sea can make just proof of the profit in them: so necessarily dependeth art and reason upon practice and experience."[80]

Such protests occasionally found their mark and elicited a reaction. Thomas Digges, himself a mathematical reformer of navigation who, unfortunately, never published the manual he was planning, became so vexed at jibes from mariners regarding his own lack of nautical experience that he actually spent fifteen weeks at sea to prove his point. His goal was to demonstrate to all concerned that the mathematically based methods he was advocating were sound in practice as well as in theory:

> In like sort by masters, pilots, and mariners, I have been answered, that my demonstrations were pretty devises: but if I had been in any sea services, I should find all these my inventions mere toys . . . adding further, that whatsoever I could in paper by demonstrations persuade, by experience on seas they found their [traditional] charts and instruments true and infallible.
>
> These constant asseverations from men of that profession, even in their own art, did make me half distrust my demonstrations, and to think that reason had abused me, or that there were some such mystery in sea service, as no land man's reason might attain unto.
>
> To resolve myself of this paradox, I spent a xv. weeks in continual sea services upon the ocean, where by proof I found, and those very masters themselves could not but confess, that experience did no less plainly discover the errors of their rules, than my demonstrations [did].[81]

Statements such as these illustrate the danger of recklessly characterizing all mathematically minded Englishmen as members of a single community of "mathematical practitioners." Norman, for example, certainly knew enough practical mathematics to qualify as one of E. G. R. Taylor's mathematical practitioners, and indeed he was included in her census accordingly. Yet he *identified himself* as a member of a community of "mechanicians and seamen," which was under attack from a different community, "the learned in the mathematicals." William Borough is also listed in Taylor's census, though he too distinguished himself from his rivals as an experienced mariner, scorning "the best learned in those sciences mathematical, without convenient practice at the sea." Thomas Digges, held up as a model mathematical practitioner by Taylor and others, found himself in dispute with a group of common mariners who had far more nautical experience than he had. Yet Digges's opponents apparently did not object to his use of mathematical methods per se, only to his insistence upon theoretical "demonstrations" when these conflicted with their own "charts and instruments true and infallible." The debate was thus over *contradictory* mathematical navigation technologies, not the overall appropriateness thereof.[82]

All of the mariners and mathematicians discussed here certainly shared at least some aspects of a common mathematical language, "tool box," and worldview; otherwise their very arguments over mathematical navigation would have been impossible.[83] Certainly they all deserve to be counted among the "mathematical practitioners" of sixteenth-century England, if such a term is to have any meaning. Yet given the highly diverse backgrounds, experiences, and skill levels of all those involved and their fundamental disagreement regarding the proper definition and practice of the art in question, can they still be usefully categorized within such a broad, sweeping "community"? Should it matter that they apparently did not perceive themselves that way? Furthermore, if not all of them should be considered part of that community, then who among them has the best claim to being the *true* "mathematical practitioners"? How much mathematics did one need to know? How much practical experience did one need to have? Was a cursory knowledge of trigonometry sufficient, or did one have to "practice" or "apply" it somehow? At what point does such a category get so malleable as to become historically meaningless?

The appeal of the term *mathematical practitioners* for the historian is undeniable. As mathematical technologies achieved a certain dominance across a wide array of arts and disciplines in early modern England (and throughout western Europe), mathematically oriented ways of perceiving and manipulating the world became ever more important and ubiquitous. Men from all walks of life, from ships' masters and instrument makers to university fellows and gentlemanly courtiers, gradually came to share many of the same instruments, methods, and approaches to solving problems that were of common interest to all of them. The trouble with gathering them all together under such a blanket category, however, is that it risks turning the *explicandum* into an *explicans*—despite its uses as an explanatory tool, it tacitly passes over a great deal that still needs to be explained. The term *mathematical practitioners* may be handy in referencing a broad spectrum of individuals who shared an interest and fluency in practical mathematics, but it inevitably obscures many of the complexities, tensions, rivalries, and cross purposes that must necessarily exist within any group or category so thoroughly diverse.

I would argue instead that historians should avoid using the term *mathematical practitioners* and resist the temptation inherent in it to treat all those to whom it might apply as any sort of organized, coherent, self-defined community. A more fruitful and less distorting approach might be to concentrate on the activities themselves, rather than on the myriad people involved with them. The arts in question, and the mathematical technologies that came to define and dominate them, might be classified together as a sort of community of disciplines, even when the individuals associated with them at all levels cannot be—early modern surveying, navigation, and gunnery can be usefully and accurately grouped together in ways that Robert Norman, Thomas Digges, and Edward Wright cannot. In discussing the "mathematical arts," therefore, the historian runs a much lower risk of overgeneralizing and thus obscuring the many rich complexities in the interactions of those who defined, practiced, and patronized them. The resulting historical picture will not only be more correct but far more interesting as well.[84]

Despite their occasional protests, virtually all of England's pilots and ships' masters did adopt at least some of the mathematical innovations to their art. As navigation became steadily more mathematized, the bar of minimal mathematical competence for practitioners rose ever higher; conservative mariners often resisted further advances by appealing to less sophisticated, "traditional" technologies that had themselves been introduced only a generation or two earlier.[85] In the seventeenth century, newer instruments such as terrestrial and celestial globes, Mercator projection charts, and John Davis's back-staff (a modified cross-staff) became standard equipment aboard English ships. As the mathematics behind the new technologies grew ever more complex, mathematicians continued to shield practitioners from the burden as much as they could. New expedients, like Edmund Gunter's navigational sector and John Napier's logarithms, removed the need for difficult calculation on the part of the mariners who used them. Once mathematical navigation technologies had been accepted by patrons and practitioners alike, mathematicians became the sole gatekeepers of navigational expertise. They would remain indispensable servants of the commonwealth for as long as they could maintain their monopoly over the vital skills and knowledge they controlled.

CHAPTER FIVE

# Francis Bacon and the Expertise of Natural Philosophy

> It may be thought indeed, that I who make such frequent mention of works and refer everything to that end, should produce some myself by way of earnest. But my course and method, as I have often clearly stated and would wish to state again, is this—not to extract works from works or experiments from experiments (as an empiric), but from works and experiments to extract causes and axioms, and again from those causes and axioms new works and experiments, as a legitimate interpreter of nature.
>
> FRANCIS BACON, *New Organon*

Francis Bacon (1561–1626) is generally regarded as one of the most important philosophers of the seventeenth century. His call for a radical reformation of the goals and methods of pursuing natural philosophy was unheralded during his own lifetime but has been widely influential ever since his death. Indeed, historians have most often tended to view Bacon and his natural philosophy in the context of what came after, associating his ideas with everything from "industrial science" and utilitarianism to the modern notion of scientific objectivity.[1] The most frequent context for historical consideration of Bacon's philosophy is probably the foundation (in 1660) and early years of the Royal Society of London. As the historian Antonio Pérez-Ramos has written, "The founding of the Royal Society represents both Bacon's deification as a philosopher and the final victory of the Baconian project of collaboration, utility, and progress in natural inquiries."[2] Despite some very real differences between the Royal Society's goals and methods and Bacon's ideas for philosophical reform, early members of the organization claimed him as their intellectual forebear. The society's first historian, Bishop Thomas Sprat, paid eloquent tribute to the importance of Bacon's works as a guiding principle for the institution:

> And of these, I shall only mention one great man, who had the true imagination of the whole extent of this enterprise, as it is now set on foot; and that is, the *Lord Bacon*. In whose books there are everywhere scattered the best arguments, that can be

produced for the defense of experimental philosophy; and the best directions, that are needful to promote it. All which he has already adorned with so much art, that if my desires could have prevailed with some excellent friends of mine, who engaged me to this work, there should have been no other preface to the *History* of the *Royal Society*, but some of his writings.[3]

As Sprat's testimony would suggest, leading members of the nascent Royal Society took Bacon's ideas very seriously; among them were Samuel Hartlib, Robert Boyle, and Robert Hooke.[4] Hooke's microscopic investigations, in particular, owed much to Bacon's philosophical method, the main elements of which he strongly endorsed in the preface to his well-known landmark work *Micrographia*.[5]

Through the Royal Society especially, Baconian philosophy has played an important role in shaping the practice of science in the West since the mid–seventeenth century. Bacon's arguments in favor of an action-oriented, collaborative, state-sponsored, institutional scientific culture were among the very earliest of their kind and have been described as his most enduring legacy to the modern Western world.[6] Yet the elements of Baconian philosophy that seem most modern did not arise ex nihilo; they were very much a product of the culture in which they were formed. It is important (if not always easy) to remember that Francis Bacon himself had been dead for thirty-four years when the Royal Society was founded and organized according to some of his philosophical principles. Bacon's own history lay firmly in the *sixteenth* century, where he lived more than half his life and where several recent historical studies have sought to locate the intellectual and social context in which his later philosophy initially took shape.[7] With respect to his proposals for reforming natural philosophy, Bacon's program shared a number of priorities and values with the Elizabethan notion of expertise; the two also had a common root in the humanist ideal of applying knowledge to useful ends. In order to understand fully the most important and influential themes in Bacon's thought, therefore, they must be situated within the sixteenth-century milieu of humanist learning, pragmatic patronage, and expertise that first gave them currency and relevance.

## Bacon and the Reform of Natural Philosophy

Bacon's vision for the reform of natural philosophy, as expressed in his later, more finished works, was a reconception of both the method according to which it should be pursued and the reasons for pursuing it.[8] He continually emphasized the ways in which he felt his approach differed most starkly and importantly from the Scholastic tradition, prevalent in most European universities. He especially condemned the Scholastic tendency to pursue philosophy through disputation and commentary, rather than through direct experience and observation of nature. According to Bacon, natural philosophy had long suffered from a misguided belief that "the dignity of the human mind is impaired by long and close intercourse with experiments and particulars . . . especially as they are laborious to search, ignoble to meditate, harsh to deliver, illiberal to practice, infinite in number, and minute in subtlety." As a result of this mistaken philosophical principle, the considerable amount of important knowledge to be gained through experience had been unwisely "rejected with disdain."[9]

Although the observation of nature was vital, it was not sufficient by itself to correct the Scholastics' many errors and shortcomings. One of Bacon's most striking ideas was that philosophers should concentrate not only on natural objects and events but also on what he termed the mechanical arts—human activities that produced or accomplished things that nature could not, or that duplicated natural actions by artificial means. This ran counter to the traditional Scholastic view of natural philosophy, derived from Aristotle, which held that mechanical activity, because it was contrived by men, was useless for the study of natural phenomena. Not only did it reveal nothing about how nature normally behaved, but it was beneath the proper social and intellectual station of the philosopher to engage in any sort of manual activity anyway.[10] Bacon argued, in contrast, that the mechanical arts offered a superior means for studying nature because they were active and productive, not sterile and disputatious, as Scholastic philosophy had become. Traditional philosophical methods, he claimed, had yielded nothing new or useful in several centuries and had in fact produced only repetitious and stagnant debates over knowledge that had been handed down from ancient Greek philosophers more than two thousand years earlier. Scholasticism was "fruitful of controversies but barren of works"; the mechanical arts were much more dynamic, "continually growing and becoming more perfect," because they were always being augmented and improved by their practitioners through daily use.[11]

Bacon's assertion of the validity and importance of human-contrived experiences as legitimate means of learning about nature was a key element in his reformed philosophical method. Whereas most Scholastic philosophers drew a strict, categorical distinction between the natural and the artificial, denying that the latter could shed any meaningful light on the former, Bacon divided the world into three related and interconnected categories: natural occurrences, natural aberrations (things and events that were not normal but were not contrived by man, e.g., "monsters"), and human artifice.[12] He argued not only that each category could yield real and valuable knowledge about nature, but that studying the monstrous and the artificial was in fact the only way to uncover natural knowledge that would otherwise remain hidden. Many aspects of nature were not easily observable under normal circumstances; only man-made experiments, Bacon wrote, would allow the diligent philosopher to remove "the mask and veil from natural objects, which are commonly concealed and obscured under the variety of shapes and external appearance."[13]

Bacon's natural philosophical reform program has been connected with the early modern fascination with *Kunstkammern*, collections of "curiosities" both natural and artificial in origin.[14] Certainly the eclectic nature of many *Kunstkammern* played a role in helping to break down the essential divide between works of nature and artifacts among natural philosophers, including as they did curiosities of all kinds, from monstrous births to mechanical automata. However, Bacon's insistence on exploring the products of human artifice in his new philosophy derived as much from pragmatic concerns as from abstract, intellectual curiosity. His inclusion of the mechanical arts as a legitimate part of the proper study of nature was rooted in his conviction that the true end of natural philosophy—which he felt most of his philosophical contemporaries had lost sight of—was to attain practical results: "Again there is another great and powerful cause why the sciences have made but little progress; which is this. It is not possible to run a course aright when the goal itself has not been rightly placed. Now the true and lawful goal of the sciences is none other than this: *that human life be endowed with new discoveries and powers.*"[15]

For Bacon, natural philosophy was as much about *using* and *controlling* nature as it was about understanding it; indeed, understanding and controlling natural knowledge were really the same thing. "Of all signs [of a true philosophy] there is none more certain or more noble than that taken from fruits," Bacon wrote. "For fruits and works are as it were sponsors and sureties for the truth of philosophies."[16] He condemned Scholasticism as sterile and barren in large part because it had failed to provide any new works to improve the lot of mankind, even over many centuries; this alone was ample evidence that the Scholastics did not have a firm grasp of how nature worked and were pursuing a flawed method. In contrast, because the mechanical arts exposed nature's hidden secrets, yielded practical results, and were themselves continually evolving and amassing new knowledge, Bacon looked to them as a fruitful means to pursue a more fundamental and accurate natural philosophy.

This is not to say that Bacon's conception of natural philosophy was crassly utilitarian, or that he did not have more intellectual and elitist goals in mind. He repeated insistently that although useful works were the hallmark of a successful philosophy, philosophical study should not be undertaken solely for the sake of practical utility. "[I]t would be an utter mistake," he wrote, "to suppose that my intention would be satisfied by a collection of experiments of arts made only with the view of thereby bringing the several arts to greater perfection."[17] Rather, Bacon believed that the mechanical arts were a valuable means to gain a better understanding of nature, and any improvements resulting therefrom were a clear sign that a better understanding had been obtained. Natural philosophy, correctly undertaken, was supposed to provide a true knowledge of the *causes* of natural phenomena. Despite the importance of the mechanical arts in the process, it was not to be the province of common craftsmen and artisans, but of learned philosophers who knew how to interpret the knowledge gained from the arts. The philosopher would not simply "extract works from works or experiments from experiments (as an empiric)" but rather would use experiments "to extract causes and axioms, and again from those causes and axioms [extract] new works and experiments, as a legitimate interpreter of nature."[18] Understanding nature's causes, Bacon argued, would *inevitably* enhance one's ability to manipulate them for practical ends. New or improved works and inventions would thus serve as the very proof and confirmation that the philosopher had at last acquired a true comprehension of nature.

The enormous project of collecting and coordinating all the myriad experiences and observations from which to construct an inductive, Baconian natural philosophy was far more than any one man, even the originator of the enterprise, could hope to manage in a single lifetime. Such an undertaking could only be "a thing of very great size . . . [such that] my own strength (if I should have no one to help me) is hardly equal to such a province." A veritable army of assistants would be required, including common craftsmen and artisans, to collect and compile all of the wide-ranging data the philosopher would need. Bacon acknowledged the necessity for a broad collaboration using the metaphor of an English trading company, sending out its mercantile factors to all the markets of the world in search of valuable commodities to import: "[T]he materials on which the intellect has to work are so widely spread, that one must employ factors and merchants to go everywhere in search of them and bring them in."[19]

Yet in order for the whole endeavor to be a success, the philosopher would have to make sure that all of his assistants carried out their respective tasks and observations correctly; otherwise their data might be unusable. This was a considerable problem, given that most of them would surely lack the intellect and subtlety of a true natural philosopher and interpreter of nature. Bacon therefore worked to set out his philosophical method in detail, so that others could follow it dependably even in his absence. Without a clear method to guide them, the legions of contributors could never collect and organize their vast multitude of natural and artificial observations into a coherent framework and derive a true philosophy from them; it would be like groping for useful objects in an endless dark space. This, Bacon believed, was precisely the problem with alchemists and other contemporary empiricists; their lack of a sufficiently structured and rigorous philosophical method reduced even their most valuable discoveries to the level of mere accidents: "But the manner of making experiments which men now use is blind and stupid. And therefore, wandering and straying as they do with no settled course, and taking counsel only from things as they fall out, they fetch a wide circuit and meet with many matters, but make little progress."[20]

With the proper method to act as a guiding principle, however, experience would serve as the only secure basis for a new natural philosophy. Moreover, Bacon believed that virtually *anyone* might aspire to be a useful contributor to the overall philosophical program, no matter what the strength or failing of his own individual intellect, if he adhered closely to the method Bacon had laid out. Following that carefully crafted method, Bacon wrote, "leaves but little to the acuteness and strength of wits, but places all wits and understandings nearly on a level." The method was therefore a vital philosophical tool with which the philosopher could build his army of contributors; it would allow even those of modest ability to accomplish things that would otherwise be difficult or impossible for them. Intelligence, it seemed, was not necessary, so long as one followed directions. The method was supposed to level men's wits in much the same way that drafting implements enabled everyone, with or without artistic ability, to draw well: "For as in the drawing of a straight line or a perfect circle, much depends on the steadiness and practice of the hand, if it be done by aim of hand only, but if with the aid of rule or compass, little or nothing; so is it exactly with my plan."[21]

The most important elements of Bacon's new approach to the study of natural philosophy, for the purposes of this chapter, were fivefold: (1) his emphasis on empirical knowledge, derived from direct observation and experience, as opposed to the endless, fruitless commentary and disputation that he vilified in the Scholastic tradition; (2) his insistence upon the essential philosophical equivalence and usefulness of natural occurrences, "monstrous" aberrations, and artificial, human-contrived experiences; (3) his focus on the importance of practical, beneficial results as the hallmark of a true understanding of nature and a successful philosophy; (4) his espousal of a large, collaborative philosophical endeavor, in which the responsibility for collecting observations and data was shared throughout a broad philosophical community; and (5) his call for adherence to a strict philosophical method, which would allow virtually anyone to contribute to the program, while structuring the accumulated practical experience into a coherent inductive philosophy.

Although not all of these elements were unique or original to Bacon, they found their earliest and most forceful exposition as a coherent, unified philosophical program when brought together in his works. They achieved their fullest articulation in Bacon's later writings—the various parts of his *Great Instauration*, published in 1620, as well as the works written shortly before (and published after) his death in 1626, including his utopian fable *New Atlantis*. However, each of these elements can also be traced back to Bacon's early intellectual career—his years at Cambridge University, his time in France as an assistant to the English ambassador Sir Amias Paulet, his legal studies at Gray's Inn, and his long and troubled quest for royal patronage during the remainder of Elizabeth's reign and the beginning of James I's. Bacon laid the foundations for his mature philosophy not in anticipation of the late seventeenth century, with which it has so often been associated, but in the midst of the late sixteenth, where it formed a part of the growing culture of English centralized government, court patronage, humanist learning, and technical expertise that has been the subject of this book.

## Bacon's Early Career

While still a young student at Cambridge, where he matriculated in 1573, Bacon had already arrived at the conviction that natural philosophy had become stagnant under Aristotelian, Scholastic dogma. As his private secretary, William Rawley, wrote of him after his death:

> Whilst he was commorant in the university, about sixteen years of age [sic], (as his lordship hath been pleased to impart unto myself), he first fell into the dislike of the philosophy of Aristotle; not for the worthlessness of the author, to whom he would ever ascribe all high attributes, but *for the unfruitfulness of the way;* being a philosophy (as his lordship used to say) only strong for disputations and contentions, *but barren of the production of works for the benefit of the life of man;* in which mind he continued to his dying day.[22]

Bacon's recollections of his university days, as he discussed them late in life with his private secretary, may well have been colored by his more recent philosophical conclusions, but this need not be assumed. Frustration with Aristotelian, Scholastic philosophy, both its method and its content, was on the rise generally during the sixteenth century, and Bacon's "dislike" was hardly unique by 1573. After all,

Peter Ramus had defended a master's thesis in 1536 at the University of Paris in which he offered a famously scathing critique of Aristotelian philosophy, natural and otherwise. In 1580 the French potter and glass-maker Bernard Palissy, who had become famous for his invention of a particular white enamel, published a negative assessment of Scholastic philosophy in his *Discours admirables*. While Bacon was resident in Paris in the late 1570s, Palissy was at the height of his renown, receiving royal patronage and giving public lectures there. It is not hard to imagine that the young Bacon would have been familiar with his work and may even have become acquainted with him, developing an even greater antipathy toward a sterile Scholasticism, as well as a deeper respect for Palissy's artisanal background.[23]

Bacon's belief in the value of mechanical/artisanal knowledge and the need for philosophy to be geared toward "the production of works for the benefit of the life of man" can be traced in large part to his experience with the humanist emphasis on practical knowledge that so pervaded English education and learned culture during Elizabeth's reign. Although Scholasticism was still a dominant force in the English university curricula, by the late sixteenth century humanism had begun to challenge its supremacy, especially in certain Cambridge colleges. The humanist curriculum was an especially popular alternative among the Elizabethan elite who sought a more practical education for their children. As the second son of Elizabeth's lord keeper of the great seal, Nicholas Bacon, Francis came from a powerful and wealthy family, and he had the same humanist educational and social background as those to whom he would eventually appeal for patronage and advancement. By 1590 humanist authors had already generated a series of highly learned treatises devoted to practical, mechanical subjects—Leon Battista Alberti and Filarete on art and architecture, Georgius Agricola and Vannoccio Biringuccio on mining and metallurgy, Taccola and Agostino Ramelli on machines and mechanics, to name but a few examples.[24] Such works were popular in England when Bacon received his education, and he was certainly familiar with at least some of them.[25]

In addition to the mechanical arts, humanism stressed the application of literary and philosophical studies to everyday life. Rhetoric, history, and moral philosophy were viewed as integral to the training of any humanist prince or magistrate, in order that he or she might draw upon these disciplines in the everyday business of government and diplomacy. With regard to private life, humanist educators taught their students that being virtuous and living a virtuous life were very much the same thing—morality had to be applied, or it did not exist. Bacon was an exemplary student of this humanist tradition, well prepared for an active career in royal administration and the law and devoted to the idea that the learned man should apply his knowledge in the service of the Crown and the commonwealth. The historian Stephen Gaukroger has argued that a key component of Bacon's natural philosophical program was his adaptation of humanist notions of practical learning and the active life to the study of nature. Like the moral philosopher, the natural philosopher must seek to apply his learning in daily life, to improve the lot of his fellow men.[26] Bacon expressed his desire to do exactly this throughout the 1580s and 1590s, as he eagerly sought to obtain patronage from his fellow humanists at the Elizabethan court.

In 1579, while he was still in Paris, Bacon received the tragic news that his father had died unexpectedly. He returned to London immediately, aware that his future prospects had been dealt a serious blow; unable to rely upon his father's position and connections to guarantee him advancement, he would have to start courting other patrons. He returned to Gray's Inn, borrowing heavily to finance his legal studies there. He sat as a member of Parliament during the 1580s and worked to ingratiate himself with various members of Elizabeth's Privy Council, all of whom he knew through his father. Although he constantly appealed to his powerful relatives and family friends, including Lord Treasurer William Burghley, who was his uncle, Bacon was repeatedly passed over for preferment. He made a few notable speeches while in parliament, but unfortunately these succeeded mainly in antagonizing the queen, further frustrating his hopes of obtaining royal patronage and advancement. Such favors as he did acquire were of little or no help to him. He was promised a clerkship in the Court of Star Chamber when one should become vacant, for example, but none did for twenty years; and although the queen had appointed him "one of her counsel learned extraordinary," this post was honorary and carried no salary.[27]

During these lean years, Bacon began to develop his first impressions of contemporary natural philosophy into a program for philosophical reform; by the early 1590s, he had incorporated this program into his bids for patronage. In 1592 he wrote a letter to Burghley, seeking some sort of preferment that would keep food on his table while also allowing him to pursue his intellectual interests. His impatience with his slight advancement thus far was clear: "I wax now somewhat ancient; one and thirty years is a great deal of sand in the hour-glass. . . . Again, the meanness of my estate doth somewhat move me: for though I cannot accuse myself that I am either prodigal or slothful, yet my health is not to spend, nor my course to get." However, even in this rather pathetic appeal to Burghley for assistance, Bacon made it clear that it was no mere clerkship he sought; he aspired to serve the queen not though modest civil service but by grand intellectual service.

"I confess that I have as vast contemplative ends, as I have moderate civil ends," he wrote, "for I have taken all knowledge to be my province."

Bacon's principal goal, he explained to Burghley, was to purge natural philosophy of "two sorts of rovers, whereof the one with frivolous disputations, confutations, and verbosities, the other with blind experiments and auricular traditions and impostures, hath committed so many spoils." In the process, he hoped to undo the damage these rovers had done and reintroduce "industrious observations, grounded conclusions, and profitable inventions and discoveries; the best state of that province," a service he regarded as nothing less than *"philanthropia."* He acknowledged, however, that he could hardly hope to undertake such a vast project by himself—this was a major reason for his quest to secure a lofty royal office, since it would give him the opportunity and resources he needed to enlist the efforts of many subordinates: "And I do easily see, that place of any reasonable countenance doth bring commandment of more wits than of a man's own; which is the thing I greatly affect."[28]

Bacon's letter to Burghley demonstrates that his ideas for the reform of natural philosophy first took shape during Elizabeth's reign. His early quest for royal patronage was meant to allow him to pursue "contemplative ends," rather than more traditional civil service per se. Twenty-eight years before the publication of the first parts of his *Great Instauration*, he had already "taken all knowledge" to be his rightful philosophical province. Moreover, the two "rovers" Bacon identified in 1592 were the very same he attacked with such vigor for the rest of his life: fruitless philosophical disputation and experience unguided by the proper method. The chief goals of his reform program likewise remained constant: "industrious observations, grounded conclusions, and profitable inventions and discoveries," all in the name of *"philanthropia"*—practical works that would benefit the commonwealth and his fellow man. Finally, Bacon recognized very early on that a program of such enormous scale would require both a large collaborative effort—the "commandment of more wits than of a man's own"—and the means to pay for it. He therefore sought to obtain a royal office "of any reasonable countenance," which would give him greater access to money and command of human resources. As early as 1592, then, the main tenets of Bacon's later philosophical works were already taking recognizable form.[29]

Bacon gave further expression to his early vision of a massive, royally sponsored program for natural philosophical research on at least two other occasions in the 1590s. In the same year that he wrote his letter to Burghley, he also composed an entertainment to celebrate the anniversary of Elizabeth's succession to the throne

(17 November), entitled "Of Tribute, Or Giving what is Due." The work was an example of the humanist genre of "mirrors for princes," in which the characters discoursed on the virtuous traits that a good prince ought to have. Bacon's "mirror," however, included an unusual element: a speech entitled "In Praise of Knowledge," in which he lauded the power of natural philosophy.[30] The speech began by asserting that knowledge pursued solely for its own sake was not enough; it had to lead to the betterment of man's estate. It was insufficient "for a man's mind to be raised above the confusion of things . . . [and] have the prospect of the order of nature and the error of men," unless he was also "able thereby to produce worthy effects, and to endow the life of man with infinite commodities." Bacon then went on to identify two barren branches of natural philosophy, both of which had failed to yield the kind of productive, operative knowledge for which he called. First came the philosophy of the ancient Greeks, which "hath the foundations in words, in ostentation, in confutation, in sects, in schools, in disputations" but "never brought to light one effect of nature before unknown." The other, more empirical tradition Bacon identified with the alchemists, who through purest chance sometimes "stumble upon somewhat which is new" but whose philosophy "hath the foundation in imposture . . . and obscurity."[31]

Neither the endless disputations of the Greek tradition nor the unguided, secretive fumblings of the alchemists would ever yield the kind of comprehensive, fruitful understanding of nature Bacon sought. They were both failures, he argued, because they fell on opposite sides of the true, productive philosophical method he advocated: "So that I know no great difference between these great philosophies, but that the one is a loud crying folly, and the other is a whispering folly. The one is gathered out of a few vulgar observations, and the other out of a few experiments of a furnace. The one never faileth to multiply words, and the other ever faileth to multiply gold."[32]

The disputatious philosophical style that the Scholastics had taken from the ancient Greek tradition and the unmethodical empiricism of the alchemists—the same two "rovers" of whom Bacon had complained to Burghley—remained the twin targets of his criticism throughout his life. In presenting this speech to the queen herself in 1592, he was certainly hoping to win from her the same kind of lofty royal appointment that he sought from Burghley, in order that he might more effectively undertake his natural philosophical reform, vanquishing the rovers in order to promote "the best state of that province" of "all knowledge."

This was not the only festive public airing Bacon gave his philosophical program. During the Gray's Inn Christmas revels (known as the *Gesta Grayorum*) of 1594–95, the students decided to transform the Inn into a pseudo-kingdom in its own right. They elected one of their number as a sort of mini-monarch, the "Prince of Purpoole," who presided over the "realm" of the Inn throughout the Yuletide festivities. The prince sat in state for the duration, even formally receiving an ambassador from the nearby Inner Temple, "their ancient allied friend." The prince was provided with his own "Privy Council" of fellow students to advise him in matters of state. Near the end of the revels, each of the six "councillors" was to deliver a public speech to the prince on a selected topic; the author of each of the councillors' speeches was almost certainly Francis Bacon. The revels were held, of course, primarily for the amusement of the denizens of Gray's Inn, but the audience for the speeches stretched far beyond the Inn's normal community. On the night of 3 January 1595, when the councillors' speeches were delivered, most of Queen Elizabeth's actual Privy Council (many of whom had had their legal training at Gray's Inn) were in attendance, including the lord treasurer, Burghley. The revels therefore provided Bacon with a splendid opportunity to promote his natural philosophy, pressing home once again his need for royal patronage to undertake it.[33]

The second of Bacon's "privy councillors" addressed the prince "advising the Study of Philosophy"; his speech may be seen as an early, skeletal framework for "Salomon's House," an idea developed much more fully in the posthumously published *New Atlantis*. All the great ancient monarchs, the councillor recited, from the Ptolemys of Egypt to Alexander the Great to Julius Caesar, had cultivated and benefited from a love of knowledge. The best and wisest course for the new Prince of Purpoole, therefore, would be to promote the study of natural philosophy as fervently as he could. Toward this end, the councillor recommended that the prince establish "four principal works and monuments of yourself." These were to include "a most perfect and general library," containing all the knowledge (ancient and modern) that had been collected in books from throughout the world; a royal garden and zoo, for the study of natural history, "so you may have in small compass a model of universal nature made private"; "a goodly huge cabinet," containing wonders of nature and also "whatsoever the hand of man by exquisite art or engine hath made rare in stuff, form, or motion"; and a "still-house," which would include furnaces, instruments, mills, and vessels, "as may be a palace fit for a philosopher's stone." This diverse collection of philosophical tools and facilities reflected Bacon's belief in giving equal philosophical weight and attention to both natural and artificial objects, and it recalls once again his desire for a large-scale, state-supported philosophical undertaking dedicated to the collection of useful knowledge.[34]

By far the longest and grandest of Bacon's early philosophical writings (and bids for royal patronage) was his book *Of the Advancement of Learning*, published in 1605 in the hope of obtaining favor from England's new King James I, who had succeeded Elizabeth in 1603 and to whom the book was dedicated. The treatise was a lengthy description of the state of knowledge and learning as Bacon saw it, divided into two books: a defense of the dignity of learning and a breakdown of its various branches and their many shortcomings. Bacon wasted no time in introducing the main elements of his by now long-formulated program for philosophical reform. In pursuing knowledge, he wrote in the first book, mankind had made many errors. "But the greatest error of all the rest," he argued, was "the mistaking or misplacing of the last or furthest end of knowledge," which was "to give a true account of their gift of reason, to the benefit and use of men . . . for the glory of the Creator and the relief of man's estate." Such an error could only be rectified "if contemplation and action may be more nearly and straitly conjoined and united together than they have been."[35] In his defense of the dignity of learning, then, Bacon provided a justification for his own particular method of learning, a method characterized both by its insistence on the need for collecting empirical observations and experiences and by its focus on producing practical benefits.

Bacon elaborated on these themes in the second book, where he divided all natural knowledge into three categories: "nature in course," "nature erring or varying," and "nature altered or wrought [by man]"—the same classifications he would later use in his 1620 essay "Parasceve." Each of these types of knowledge, in turn, corresponded to a specific branch of philosophical investigation: "history of Creatures, history of Marvels, and history of Arts."[36] With regard to the latter branch, Bacon lamented that so many philosophers still thought it beneath them to take a more active, empirical approach: "I find some collections made of agriculture, and likewise of manual arts; but commonly with a rejection of experiments familiar and vulgar. For it is esteemed a kind of dishonour unto learning to descend to inquiry or meditation upon matters mechanical."[37]

The contemporary ignorance and disdain for the "History of Nature Wrought or Mechanical" was unacceptable to Bacon, for whom this class of knowledge represented potentially the most substantive and promising field of philosophical endeavor: "But if my judgment be of any weight, the use of History Mechanical is of all others the most radical and fundamental towards natural philosophy; such natural philosophy as shall not vanish in the fume of subtle, sublime, or delectable speculation, but such as shall be operative to the endowment and benefit of man's life." A full understanding of nature could be acquired only by incorporating the study of human action upon natural objects, he argued, because nature could be constrained thereby to yield secrets that would otherwise remain hidden: "For like as a man's disposition is never well known till he be crossed, nor Proteus ever changed shapes till he was straitened and held fast; so the passages and variations of nature cannot appear so fully in the liberty of nature, as in the trials and vexations of art." The knowledge gained through human-contrived experiments, in turn, would "not only minister and suggest for the present many ingenious practices in all trades," but more importantly, "it will give a more true and real illumination concerning causes and axioms than is hitherto attained."[38] Thus the best sign of a true natural philosophy would be the practical benefits that it would inevitably yield.

The core elements of Bacon's later philosophical works were thus almost fully formulated in his mind at least fifteen years before they took mature form in his *Great Instauration*. Moreover, they surfaced repeatedly in his numerous attempts to obtain royal patronage during the Elizabethan and early Jacobean reigns. Bacon knew that if his program for philosophical reform was ever to come to fruition, it would need to have the backing of wealthy, powerful patrons who could grant him the resources and freedom to pursue it. The fact that Bacon obviously believed his ideas would be attractive enough to Elizabethan courtiers to deserve their material and intellectual support says much about his perceptions of English patronage priorities at the end of the sixteenth century. With his emphasis on the importance of empirical observations, guided by a proper theory or method and leading ultimately to new and improved practical results, Bacon was appealing directly to his patrons' humanist-inspired focus on active, applied knowledge—a focus shared by nearly all the English ruling elite, himself included. He was not the only one to try such a strategy; whether consciously or unconsciously, Bacon's natural philosophy shares a number of goals and methods in common with contemporary works produced by the growing ranks of English technical experts. Their myriad instructional manuals, after all, were intended to win them patronage from precisely the same people whom Bacon was courting; it is unsurprising that they should have adopted a similar approach to obtain it. To understand the origins of Bacon's natural philosophy, therefore, it must be situated within the Elizabethan culture of expertise where he first formulated it.

## Bacon and His Expert Contemporaries

Bacon's early philosophical works, and his use of them in his quest for patronage during the 1590s, may be evaluated within the broader sixteenth-century culture of technical expertise by looking again at some of the navigation manuals and practices discussed in the previous two chapters. All of the manuals in question were published during Bacon's lifetime, with many of the most important and popular examples circulating at exactly the same time that Bacon was studying at Gray's Inn and trying to gain advancement at court. While Bacon wrote his festive orations on the value of empirical knowledge and the inevitable practical benefits of a sound natural philosophy, mathematicians such as Thomas Hood, William Blundeville, Edward Wright, and Thomas Harriot were lecturing and tutoring the ruling elite of Elizabethan London on the practical value of mathematical navigation. These men may not have been looking for quite the same level of patronage that Bacon sought, but they still had to make their skills and services appealing to the same set of patrons, and they used similar means to do so.

The English navigation manuals constituted an established and thriving genre of literature. With all of their many editions, dozens of different treatises dedicated to mathematical navigation had been published by 1600. The desire of so many authors to share their knowledge with mariners, mathematicians, patrons, and all other readers may be seen as more than just self-advertisement; it also indicates the perceived value of intellectual collaboration within the culture of expertise. Indeed, the manuals' authors sometimes made explicit reference to the works of their colleagues and rivals and acknowledged building upon them. William Bourne, for example, wrote that his *Regiment for the Sea* contained "other necessary things meet to be known in navigation, and not mentioned in the book of Martín Cortés called the *Arte of Navigation*." He did not feel the need to cover material already addressed by Cortés, "for that it is there sufficiently declared already."[39] Thomas Blundeville described the final "Exercise" in his tome as containing "the first and chiefest principles of navigation . . . lately collected out of the best modern writers, and treaters of that art," though he did claim that his own version was "more orderly taught" than some previous attempts.[40] At least some authors of navigation manuals believed that understanding and improving even the relatively humble art of navigation was a collaborative endeavor, larger than the efforts of any single contributor could encompass.

Navigational instructors such as Sebastian Cabot also urged practical collaboration among their protégés, to help ensure that an individual's mistakes were caught early, before they could cause any serious harm. Cabot taught his newly trained pilots to work together in plotting their courses because he recognized that the task was too important and complex to be wholly entrusted to any single, inexperienced man.[41] Accurate navigational data were important not only for the safety of the ships and crews directly relying upon them, but also for the future use of the company's mariners, who had to be able to retrace their steps in order to open and maintain a regular trade with a new market. English geographers such as William Borough and John Dee also recognized the necessity of collaboration. In attempting to assemble a comprehensive and accurate body of geographic and navigational knowledge from around the globe, they understood that the successful completion of their project would require the joint contributions of a host of mariners, since no one man would ever be able to collect the multitude of observations they needed.[42] Within the arts of navigation and cartography, collaboration served as a means both of avoiding individual errors and of taking on a project of larger scale than anyone could manage alone. This recalls Bacon's proposals to reform natural philosophy, in which he praised the manual arts for their high degree of collaboration and corroboration, allowing them to be perfected over time through the accumulated efforts of innumerable contributors. He hoped that the same would ultimately come to be true of natural philosophy as well.[43]

Bacon was not the first to ask wealthy and powerful patrons to support and endow a new institution to foster large philosophical undertakings and improve the practical arts. Stephen Borough's bid to create and fund a new office for a chief pilot of England was a bold attempt to enlist the authority and resources of the Crown in the effort to teach English mariners the mathematical foundations of their art.[44] Although Borough's petition never received the Crown's full and overt support during his lifetime, later attempts by other petitioners were somewhat more successful. Thomas Hood's London mathematical lectures were mandated by the Privy Council in 1588 and paid for by the city of London, in order that England's militia and naval officers would be better prepared to meet the mathematical challenges they would face in their posts. In 1590 Hood successfully appealed to the Privy Council to extend its support of his lectures for another two years, though he failed to win a permanent endowment for the institution.[45] Given the contemporary existence of an established, institutionalized, applied mathematical lectureship in London, Bacon's proposal to reform the whole of natural philosophy and reorient it toward practical ends might have been made to seem like an even better investment to the pragmatic patrons thereof.

Although many of the navigation manuals' authors were not themselves practicing navigators, nearly all of them perceived the importance of either claiming to have practical experience or, at least, praising it in their introductions. The landlubber Thomas Hood wrote that he "had to do a long time with diverse of your profession [mariners], both for the making of sea cards; and also for instructing them in mathematical matters belonging to navigation." It was this experience, in fact, that had inspired him to teach mathematical navigation in the first place, for he had "found many [mariners] willing to learn, and by that means had an insight into their wants." He also hoped that his instruction would "be not prejudicial to any other man's, whose experience in hydrographical matters is more than mine."[46] John Davis, a man with impeccable nautical credentials himself, was quick to praise English mathematicians such as Thomas Digges, John Dee, and Thomas Harriot "for theorical speculations and most cunning calculation." Yet in the same breath, he paid equal tribute to the master shipwright Matthew Baker, "for his skill and surpassing grounded knowledge for the building of ships advantageable to all purpose," and to several English maritime explorers, whose combined exploits proved that "not only in the skill of navigation, but also in the mechanical execution of the practices of sailing, we are not to be matched by any nation of the earth."[47] William Borough criticized his land-bound rivals, claiming that "none of the best learned in those sciences mathematical, without convenient practice at the sea can make just proof of the profit in them: so necessarily dependeth art and reason upon practice and experience."[48] And Thomas Digges defended himself from criticism by claiming that he had personally "spent a xv. weeks in continual sea services upon the ocean, where by proof I found, and those very [ship's] masters themselves could not but confess, that experience did no less plainly discover the errors of their rules, than my demonstrations [did]."[49] Expert navigators, it seems, were ideally supposed to get their feet wet at sea—just as true Baconian philosophers were supposed to get their hands dirty in investigating nature for themselves.

There was no doubt where all of this enlightening experience at sea was supposed to lead: virtually every author claimed to have written his manual for practical reasons, to correct or improve upon current navigational practice. Edward Wright called his treatise *Certaine Errors in Navigation,* implying that his book would correct and eliminate the same, whereas the organization of Thomas Blundeville's *Exercises* led inexorably from simple arithmetic to the practical, nautical fruits ultimately derived from it. If Bacon could complain that Scholastic philosophers scorned experience and disdained fruitful endeavors, the same could not be said of contemporary experts in the mathematized art of navigation. The mathematicians' repeated declarations of the practical benefits of their knowledge were meant to appeal to wealthy patrons who invested heavily in overseas trading ventures and were also principal consumers of the books, lectures, and instruments the mathematicians produced. The widespread value that humanist-educated Elizabethans placed on practical knowledge likewise prompted Bacon to include a thorough understanding of the human, mechanical arts as a core component of his natural philosophical program.

Yet for the authors of English navigation manuals, as for Bacon, experience and practical results were still not enough by themselves. The true expert in the art of navigation should also have mastered the theoretical principles behind his art; that is to say, his experiences and actions aboard ship had to be guided and ordered by a more abstract, methodical approach. Without this deeper, mathematical understanding, the practitioner was still just an empiric groping about in the dark, and any success he might achieve thereby was primarily the result of good luck. As William Bourne wrote, "[W]hatsoever he be that will say that he can do any thing, and *if that he cannot show the reason of the doing thereof* I do say unto you he cannot do it, and this is most certain for if that he doth it, he doth it but by fortune."[50] According to William Borough, a properly trained practitioner should be able "not only [to] judge of instruments, rules, and precepts given by other[s], but also be able to correct them, and to devise new of himself. And this not only in navigation, but in all mechanical sciences."[51] This could only happen when the practitioner possessed a true, theoretical comprehension of his art. Edward Wright believed that such an understanding was so superior to mere experience in establishing true expertise that the art of navigation could "by him that hath been but meanly conversant in mathematical meditations be better apprehended, than otherwise it can by the seafaring man, though he spend his whole life in sailing over all the seas in the world."[52] Bacon's abiding belief in the importance of an overarching method as a prerequisite for fruitful philosophical endeavor is not far removed from these authors' emphasis on theoretical understanding as a necessary guide for successful practice.[53] The unlearned practitioner, nautical or philosophical, might subsist in either case, but he could never aspire to true mastery of his discipline unless he comprehended and followed the fundamental principles underlying his practice.

With respect to the practical applications of knowledge, the ideal of collaboration toward a cumulative goal, the role of royal patronage in supporting such an endeavor, and the insistence upon a coherent theory or method as a guide to proper practice, Francis Bacon's natural philosophy incorporated many key elements of the Elizabethan culture of expertise. This is not to say that Bacon deliberately modeled his proposals for philosophical reform on the navigation manuals considered above, or that he and their authors were in direct competition for patronage. Rather, the similarities between the various works indicate that he was attempting to obtain patronage in the same cultural milieu, one that valued expertise and the greater control over useful knowledge that it purported to provide. Indeed, Bacon himself shared the same humanist education and outlook that had led Elizabethan patrons to privilege practical knowledge in the first place.

Baconian natural philosophy might even be interpreted as a different manifestation of early modern expertise, with Bacon cast as a sort of philosophical expert mediator. Much like his fellow experts, Bacon appealed to patrons by offering to put his valuable skills and knowledge at their disposal. His reforms were intended to place natural philosophy and its practitioners under the control of the Crown and court, in order that their discoveries might be more effectively exploited by the patrons who sponsored their work. He even applied to natural philosophy one of the consummate techniques of expert mediation: black-boxing. Just as the mathematicians black-boxed their new navigational technologies to make them easier for nonexperts to use, Bacon reserved the most difficult aspects of natural philosophical inquiry to himself, leaving to everyone else the comparatively minor roles of collecting data and applying the fruits of the knowledge that he alone could provide.

## The Black-Boxing of Natural Philosophy

The authors of sixteenth-century navigation manuals used their superior knowledge of mathematics to render the art of navigation ever more mathematical. At the same time, they presented their new technologies in such a way as to conceal as much of the difficult mathematics from their readers as they could, completing the calculations themselves and then black-boxing them through the use of tables and other instruments. The reader-practitioner was then reduced to a basic input-output function; he needed little real understanding of how the technologies in question actually worked. In much the same way, Bacon hoped to press an army of observers and experimenters into service, having them collect the vast amounts of empirical data he would need in order to build his new, inductive natural philosophy. Toward that end, he provided all the instructions the average educated man would need in order to follow his all-important philosophical method properly. Yet Bacon assumed full responsibility himself for the method's most intellectually challenging step: abstracting nature's axioms and laws, the very essence of the new philosophy, from the masses of raw data to be collected by his reader-practitioners. In other words, Bacon black-boxed the most critical and demanding phases of his method, so that even those of modest ability could both contribute to it and benefit from it, while he alone comprehended its deeper, more important foundations and ends. He made himself the sole, indispensable expert of his own natural philosophy.

One of the best-known of Bacon's philosophical aphorisms, published in his *New Organon*, dealt with his famous "leveling of wits," in which he claimed that his reformed philosophical method was so pure and basic in its reliance on the collection of empirical data that virtually anyone could play an active role. One great advantage of his method, Bacon wrote, was that it "leaves but little to the acuteness and strength of wits, but places all wits and understandings nearly on a level." All contributors would have an assigned role to play in constructing the new philosophy, one best suited to their own interests and abilities. In order for the whole system to work, however, it was vital that all contributors adhere strictly to the Baconian method, whatever their role within it might be. This would help to eliminate individual variances of judgment and skill, allowing philosophical knowledge to advance most efficiently, with none of the mistakes, dead ends, or duplicated efforts that had plagued other philosophies: "For as the saying is, the lame man who keeps the right road outstrips the runner who takes a wrong one."[54] Bacon's method, then, appeared to open the doors of philosophical inquiry to anyone who wished to participate in it—a necessary tactic, he knew, if he hoped to achieve a full reformation of natural philosophy on the scale he intended. His new philosophy would be a collaborative success, or no success at all.

This apparent move to democratize natural philosophy was a false one, however; Bacon's proposed philosophical method was in fact rigidly hierarchical, with himself at the top of the pyramid. This should not seem surprising, given the deeply stratified nature of early modern English society and the lofty station Bacon's family enjoyed within it. Sir Francis Bacon, Baron of Verulam, was certainly no man of the people; like his father before him, he was one of the wealthiest and most powerful men in the realm, very much a member of the ruling elite. The historians Julian Martin and John E. Leary have shown that, despite the posthumous attention given to his appeals for a broad philosophical collaboration, Bacon's natural philosophy was every bit as exclusive and hierarchical as his political thoughts and legal opinions were. In his contemporaneous proposals for reforming England's convoluted common law system, for example, Bacon's principal intent was to reserve the burden of interpreting the law to a small, select committee of legal experts who answered directly to the Crown, thus taking away the most powerful role of the English judiciary.[55] In much the same way, Bacon envisioned a small cadre of natural philosophers sifting through the masses of data collected by less skilled practitioners and using the data to derive inductively the axioms and laws—too subtle for others to see—of the new philosophy. As natural philosophical experts, they alone could hope to speak with the voice of nature itself.[56]

As early as 1603, Bacon had already placed himself and his few intellectual peers well above the basic and limited contributions of the common philosophical practitioner. His philosophy was to be collaborative, but segregated. For one thing, not all parts of the philosophy were to be shared throughout the broader philosophical community. In his drafted preface to an intended but unwritten treatise called "Of the Interpretation of Nature," Bacon wrote that selective secrecy would limit the dissemination of both his method and its results. Although the preliminary portions of the new philosophy might "be published to the world and circulate from mouth to mouth," the later, more important parts should be restricted and "passed from hand to hand, with selection and judgment." In answer to critics who might claim that such secrecy was "an old trick of impostors" designed to keep the masses enthralled, Bacon replied that his project could only prosper if left in the exclusive care of those skilled, responsible philosophers best able to advance it. "[I]t is no imposture at all," he wrote, "but a sober foresight, which tells me that the formula itself of interpretation, and the discoveries made by the same, will thrive better if committed to the charge of some fit and selected minds, and kept private."[57] This was very far indeed from a democratization of knowledge; the most important elements of the entire philosophical program—the interpretation of data into natural axioms and laws and their subsequent translation into fruitful works—were reserved for a select few. Bacon's reformed philosophy was to be the province of the intellectual elite; the common practitioners—those presumably collecting all of the observations needed to make his interpretations—were to be excluded altogether from the more subtle and rarefied phases of the endeavor.

Bacon did not see fit to revise his elitist program over time, either. In a short essay of 1620 entitled "Parasceve," he outlined his inductive method for compiling "a natural and experimental history" and appended it to his *New Organon*. In the essay, Bacon explicitly placed himself above the mean labors of those actually performing the experiments and supplying the observations for his philosophy. He declared that it would be "somewhat beneath the dignity of an undertaking like mine" for him to be directly involved with the actual observations, "a matter which is open to almost every man's industry." Instead, he reserved for himself the most demanding parts of the endeavor, which he considered a more appropriate use of his superior philosophical skills: "That however which is the main part of the matter I will myself now supply, by diligently and exactly setting forth the method and description of a history of this kind . . . lest men for want of warning set to work the wrong way . . . and so go far astray from my design."[58] Not that Bacon was unmindful of the important role played by the legions of practitioners working under him; without their participation, after all, his philosophy would come to nothing. Yet he always held his own contribution—the philosophical method according to which the observations were to be made—as the cornerstone of the whole endeavor, the piece of the puzzle that only someone of his intelligence and talent could provide. With his multitudes of subordinates taking his prescribed method as their guide, "the investigation of nature and of all sciences will be the work of a few years." Without his method, however, all would be confusion and chaos, and real progress would be impossible: "[I]f all the wits of all the ages had met or shall hereafter meet together; if the whole human race had applied or shall hereafter apply themselves to philosophy, and the whole earth had been or shall be nothing but academies and colleges and schools of learned men; still without a natural and experimental history such as I am going to prescribe, no progress worthy of the human race could have been made or can be made in philosophy and the sciences."[59] Francis Bacon, the creator of this new, potent philosophical method, was also to be the sole expert in its full implementation.

The social and intellectual elitism of Bacon's proposals for natural philosophical reform are perhaps most evident in his utopian fable, *New Atlantis*, written near the end of his life and published posthumously in 1627. In this vision of an ideal society, not only was natural philosophy devoted to the service of the state, but the philosophers who controlled it apparently also constituted one of the highest civil authorities in the realm. The philosophers themselves were organized into a strict hierarchy, with a very small number at the top dominating the collection and interpretation of information, revealing the fruits of their labor to the rest of the populace as they saw fit. In short, Bacon wrote *New Atlantis* to illustrate how natural philosophy could become a powerful tool in a centralized, hierarchical, imperial monarchy, an awesome manifestation of the majesty of the king, once the philosophers were granted the royal support and patronage that was their due.[60]

Bacon's narrator in *New Atlantis* was a Spanish mariner whose imperiled ship had accidentally found a small but incredibly prosperous island in the Pacific Ocean, called Bensalem by its inhabitants. With many crewmen ill and a dwindling supply of food, the crew of the narrator's ship were permitted by the inhabitants to land at Bensalem for relief. Once on shore, they were lodged in the "Strangers' House," fed with all manner of delicious food and drink, "better than any collegiate diet that I have known in Europe," and nursed back to health by means of some miraculous-seeming medicines.[61] In time, as the crew gained the confidence of the island's Christian inhabitants, the governor of the Strangers' House came to see them. During his visit, he offered to answer some of their questions about the island and its people, after which they were permitted to explore a bit on their own.

Central to the island's prosperity, the governor explained, was an ancient institution of natural philosophy called "Salomon's House," named for the "King of the Hebrews" and "dedicated to the study of the works and creatures of God." Salomon's House was established "for the finding out of the true nature of all things, (whereby God might have the more glory in the workmanship of them, and men the more fruit in the use of them)." The institution's mission thus echoed the "furthest end of knowledge" as Bacon expressed it twenty years earlier in his *Of the Advancement of Learning:* "to give a true account of [man's] gift of reason, to the benefit and use of men . . . for the glory of the Creator and the relief of man's estate" (3:145–146).[62] Salomon's House had obviously generated many benefits and marvels for the good of the island's inhabitants; in their subsequent explorations, the sailors were "continually . . . met with many things right worthy of observation and relation; as indeed, if there be a mirror in the world worthy to hold men's eyes, it is that country." After the crew had a chance to get to know a number of their Bensalemite hosts, one of the small group of eminent philosophers who ran Salomon's House, known as "the Fathers," paid a visit to the city. The arrival of the Father was no ordinary or minor event in the lives of the city dwellers; as the narrator was informed, "His coming is in state; but the cause of his coming is secret," and there had been no such visit there for at least a dozen years (3:147, 154).

The Father entered the city in a most elaborately decorated horse-drawn chariot, with two footmen on either side, preceded by fifty attendants and followed by "all the officers and principals of the Companies of the City," recalling the grand spectacle of a royal procession through London. He was dressed in a colorful array of linens, velvets, and silks. The whole city turned out to welcome him, though in a most seemly and orderly fashion: "The street was wonderfully well kept: so that there was never any army had their men stand in better battle-array, than the people stood. The windows likewise were not crowded, but everyone stood in them as if they had been placed" (3:154–155). If Bacon intended Salomon's House to be the utopian embodiment of his philosophical reform program, then the Father of Salomon's House was a personification of the elitism inherent in that program. It would be difficult to imagine a more regal entry or enthusiastic welcome, particularly for a figure who was not even the monarch or governor of the island. Moreover, the reason for his visit to the city was carefully kept secret from the inhabitants, and indeed it was never revealed in the narrative, further enhancing his mysterious authority. The Father was easily the most powerful figure portrayed in the fable, entirely because of his exclusive control of the natural philosophical knowledge the Bensalemites valued so highly.

Three days after the Father's entrance to the city, the narrator was granted a rare privilege: he was to be admitted to the Father's presence, to have a "private conference" with him. On the appointed day, the Father ordered even the pages from the room, before he began to speak to the narrator in fluent Spanish: "God bless thee, my son; I will give thee the greatest jewel I have. For I will impart unto thee, for the love of God and men, a relation of the true state of Salomon's House." First the Father revealed the House's mission: "The end of our foundation is the knowledge of causes, and secret motions of things; and the enlarging of the bounds of human empire, to the effecting of all things possible" (3:155–156). He then went on to describe at great length all of the "preparations and instruments" used in the House's explorations of nature, saying at least as much about the marvelous works derived from their investigations as he said about the tools and experiments themselves. Much like Bacon in his 1592 letter to Burghley, the Fathers of Salomon's House took all knowledge for their province, having erected experimental facilities in caves deep underground, on top of the highest mountains, and everywhere in between. They owned elaborate towers, lakes of fresh and salt water, artificial baths, orchards and gardens, brew- and bake-houses, houses dedicated to experimenting with each of the human senses, and engine houses to experiment with motion (3:156–164). In short, Salomon's House was responsible for the construction, operation, and maintenance of an enormous natural philosophical apparatus, with countless instruments and innumerable (though usually unmentioned) subordinate technicians to run them, making observations and reporting their findings directly to the Fathers. It was the ultimate manifestation of Bacon's institutionalized, collaborative, inductive natural philosophy.

It was also the ultimate expression of Bacon's philosophical elitism. How many Fathers of Salomon's House, one might wonder, were needed to manage and oversee such a vast, comprehensive, and labor-intensive operation? Precisely three dozen, not counting a small number of "novices and apprentices" employed so that "the succession of [the Fathers] do not fail." The thirty-six Fathers were powerful, wealthy, and highly honored for their work. The entrance of just one Father into the city provided the reader with an ample display of their opulence and the esteem in which they were held by the Bensalemites. All of the Fathers occasionally made such "circuits or visits of diverse principal cities of the kingdom," during which they revealed new inventions and discoveries and gave advance warning of coming plagues, tempests, comets, and other important events. Presumably, then, regal entrances and civic honors were a regular part of the Fathers' lives, even if their visits to any particular city were relatively rare. Most strikingly, although the Fathers of Salomon's House were not the secular rulers of the island, in some ways they were even more powerful, for they decided among themselves which discoveries to share and which to keep secret: "And this we do also: we have consultations, which of the inventions and experiences which we have discovered shall be published, and which not: and take all an oath of secrecy, for the concealing of those which we think fit to keep secret: *though some of those we do reveal sometimes to the state, and some not*" (3:165–166, emphasis added). Even the king and his ministers, it appears, were not privy to the entire body of natural knowledge and operative power controlled by the Fathers of Salomon's House.

Nor was philosophical elitism imposed only upon the uninitiated; even within the walls of Salomon's House, the thirty-six exalted Fathers were strictly hierarchical in their own division of labor. At the bottom rung of the ladder, Bacon placed the twelve "Merchants of Light," who secretly traveled to foreign countries, searching out the new and useful knowledge of other peoples; then came the three "Depredators," who collected "the experiments which are in all books"; the three "Mystery-Men," who gathered experiments from the mechanical and liberal arts; and the three "Pioneers" or "Miners," who tried new experiments themselves (3:164). These twenty-one Fathers functioned as the managers of the unnamed multitudes at work throughout the island, performing experiments, making observations, and reporting their findings to Salomon's House.

As in Bacon's other works, the interpreters of natural philosophical knowledge enjoyed a higher status than mere collectors of information. Ranked just above the Fathers responsible for conducting experiments and gathering data were the three "Compilers," who "draw the experiments of the former four [classes] into titles and tables, to give the better light for the drawing of observations and axioms out of them." Then came three "Dowry-Men" or "Benefactors," who combed through the tables produced by the Compilers and tried to draw out clear demonstrations and useful applications from them. Above these were the three "Lamps," who devised new experiments that would provide answers to particularly crucial or challenging questions. Then came three "Inoculators," who undertook to perform the special experiments mandated by the Lamps. Finally, at the pinnacle of philosophical power and prestige, "we have three that raise the former discoveries by experiments into greater observations, axioms, and aphorisms. These we call Interpreters of Nature" (3:164–165).

Within the elite ranks of the thirty-six Fathers, then, there was an ultra-elite corps of three Interpreter-Fathers, who alone were ultimately responsible for shaping and finishing the new philosophy from the parts generated and organized by all of their subordinates. This system represented expert mediation on a whole new scale. As an obligatory point of passage for all natural knowledge, the Interpreters were the only individuals in Salomon's House, and thus in all of Bensalem, who comprehended every aspect of natural philosophy. All others, though they varied widely in their respective roles, capabilities, and degree of privilege, were more or less reduced to an input-output function. They collected data and passed it up the ladder to their philosophical superiors, gratefully accepting whatever practical benefits were handed down to them.

The whole of natural philosophy, in other words, had been effectively black-boxed by those few at the top of the system. Only the Interpreters had full control over access to natural knowledge, to the practical benefits deriving from it, and most importantly, to the entire means of producing and interpreting that knowledge. The traditional secrecy of craft knowledge had been completely inverted; whereas craft practitioners in medieval and early modern Europe had once protected their valuable knowledge by keeping their techniques secret, in Bensalem it was the natural philosophers who enjoyed exclusive access to all valuable knowledge and could place it at the disposal of the monarch and commonwealth. As consummate expert mediators, the Interpreters dominated the myriad practitioners who worked beneath them and served their royal patrons as the only true arbiters of natural philosophy. Bacon, as the creator of the method according to which the whole system operated and the sole expert in its full implementation, would undoubtedly have placed himself in their small number.

When he had finished his disquisition, the Father of Salomon's House arose and gave his blessing to the narrator, granting him permission to make public all that the Father had said, for the good of all men throughout the world. Then, with a parting largesse of two thousand ducats, he left (3:166). The treatise ends abruptly at this point. The narrator was not suffered to ask the Father any questions; indeed, he spoke not a single word during the "conference." William Rawley, who was responsible for editing and publishing the treatise after Bacon's death, wrote in his own preface that Bacon had intended to add "a frame of laws [for Bensalem], or of the best state or mould of a commonwealth" (3:127).[63] It is most regrettable that he did not finish this projected extension of the work, wherein the reader might have had a glimpse of Bacon's utopian Privy Council, in their roles as the most generous patrons of philosophical inquiry and most grateful consumers of the fruits thereof. Nevertheless, in his unfinished *New Atlantis*, Bacon presented an idealized vision of his elite, state-sponsored philosophical enterprise, organized according to the new philosophical method he had spent his entire adult life refining and advocating and in which he would be the principal and undoubted expert.

To claim that Francis Bacon's program for the reform of natural philosophy was the wholesale product of an English culture of expertise that came into prominence during the first half of his life has not been my intent. Although Bacon frequently expressed his respect for those who had personal experience in exploring nature or possessed practical skills and knowledge, he did not distinguish any who might be called experts among them, and he certainly did not view them as his intellectual creditors. Nevertheless, the values and priorities of Bacon's prospective patrons—their humanist-inspired demand for practical, action-oriented knowledge that could help them extend their control over the world they lived in—did influence

Bacon's structuring of his natural philosophy. Indeed, their values were also his own; his family's wealth, social status, and political connections ensured that Bacon was educated from early childhood to share the appreciation for practical knowledge common throughout the Elizabethan ruling elite. At the same time, while he was still a young man studying in London, Bacon was deeply immersed in a society where experts were beginning to proliferate and prosper. He certainly walked the same streets as they, knew many of the same people at court, and had a great respect for the valuable knowledge they possessed.[64]

After concluding that the natural philosophy of his time was fatally stunted and had little to offer humanity, Bacon faced the daunting task of finding something with which to replace it. As a young, ambitious man looking to make his way at court, he also had to appeal to patrons who could advance has career. Given that more and more Elizabethan technical experts were successfully gaining the favor of those same patrons, it is not altogether surprising that Bacon should have adopted some of their strategies and general outlook in formulating his own natural philosophical method. The experts offered their patrons greater control over the useful knowledge they craved; Bacon, however, believed that he could give them something more: control over the entire process of generating and interpreting that knowledge. Just as the experts helped their patrons to govern and manage an increasingly powerful and centralized realm, Bacon self-consciously envisioned his reformed philosophy as a fundamental part of an imperial monarchy.[65] The enthusiastic reception and adaptation of his ideas after his death, both in England and throughout Europe, led to important consequences for the development of modern science, technology, and statehood. Any attempt to understand these historical trends must begin by situating Francis Bacon and his philosophy where they originated: the Elizabethan culture of humanism, patronage, and expertise.

CONCLUSION

# Power, Authority, and the Expert Mediator

In the preceding chapters, I have attempted to trace the origins and describe the role of a new figure in early modern England: the expert mediator. By the end of the sixteenth century, the technical expert had done much to distinguish himself from the common practitioners who labored at the tasks in which he asserted his expertise, and he was seen by many of his contemporaries as superior to those workers. He succeeded in this because the very notion of expertise was in flux throughout the century, shifting from an emphasis on personal experience to the possession of a more theoretical kind of knowledge and skill. Traditional practitioners, the new experts asserted, might know *how to do things*, but their knowledge was inherently local, circumstantial, and limited. Experts, by contrast, comprehended *how and why things worked*, which represented a different class of knowledge altogether. Experts therefore laid claim to greater intellectual mobility and flexibility than mere practitioners could ever possess; their expertise was supposed to be more broadly applicable, across a wider variety of situations. To be an expert was not simply to know how to solve a given problem, but to be able to anticipate innumerable related problems, approaching their solution in ways not readily available to most practitioners.

Experts were one manifestation of a particularly pragmatic, action-oriented branch of Renaissance humanism, especially popular among the educated Elizabethan elite, which emphasized the importance of applying one's knowledge to practical ends. They were frequently men of considerable learning themselves;

some were self-educated, but many had spent time at university, and several had even obtained fellowships there. In order to place themselves on a higher intellectual and social footing than common practitioners, they made use of their education, relying on their more theoretical, skill-based approach as a reason for and justification of their enhanced status as experts. Mathematics was particularly serviceable in this regard because it provided a ready means whereby the basic tenets of some highly valued arts could be redefined and abstracted into theoretical principles. This allowed age-old problems to be perceived in a new way and generated solutions that only the experts themselves were fully competent to control. Many experts even published instructional manuals, as an open advertisement to potential patrons of their superior command of the arts in question. Experts were also usually quite at ease in gentlemanly and noble circles; some of them were members of the landowning classes themselves, while others came from clerical or wealthy mercantile backgrounds, or at the very least were eminent craftsmen, accustomed to dealing with a clientele from among the "better sort." They tended to be accepted and trusted by gentlemen and wealthy patrons in ways that most craftsmen and practitioners never could have been. The experts' claims of a wider applicability and higher value for their knowledge, their generally more advanced and refined education, their greater comfort in elevated social circles, and their very public, literary portrayal of themselves as the superior masters of their arts all helped to make them seem more attractive as clients to prospective patrons.

Experts came into being and prospered in part because they helped to meet a growing administrative demand in Elizabethan England, and indeed throughout all the increasingly centralized, corporate, and militarily powerful nation-states of western Europe. As the Tudor Privy Council gradually established its authority to govern throughout the entire realm, the councillors came to be involved directly in the management of what had traditionally been regarded as local affairs. The farther from London they sought to exert their administrative will, the thinner they spread themselves, and the more local toes they trod upon. To govern the more peripheral areas of the kingdom most effectively, it became increasingly important for them to maintain trusted subordinates at the local level, persons who could serve as their eyes, ears, mouths, and hands. These royal clients acted very much as mediators, representing the Crown's authority to the locality and communicating local affairs and concerns to the Privy Council. Experts were thus one example of a broader management technique, useful for controlling situations in which technology and technical knowledge played a major role.

In the management of large-scale, expensive, and technically challenging undertakings throughout the realm, expert mediators were potentially of enormous value because they could help to keep their patrons connected to the projects in question, providing the crucial skills and knowledge that the patrons themselves lacked. The expert mediator ideally enabled his patrons to understand what was going on with respect to a given project, to make reasoned decisions about what should happen next, and to see to it that those decisions were implemented on the ground. In placing his knowledge at the disposal of those who relied upon him, the expert thereby gave his patrons effective control over his own field of expertise. The strategy was not always successful; when clients did not actually possess the sort of expertise they claimed, like Thomas Thurland at the Keswick copper mines or John Trew at Dover Harbor, they could neither provide their patrons with the enhanced control they sought nor mediate successfully between the center and the periphery. When the clients' claim to expertise was convincing to patrons, however, as Thomas Digges's ultimately was in managing the Dover Harbor project, they gained the trust of those they served and played a key administrative role as mediators in the newly centralized state. In short, expert mediators gave their patrons greater control not just over local administration, but over local *knowledge* as well, knowledge that had previously been restricted to the secretive world of craft practitioners.[1]

Because the experts alone could claim to possess and control the kind of knowledge that would be of greatest use to their patrons, they soon became indispensable as clients. Common practitioners, it seemed, could not hope to manage complex projects or bodies of information with the same degree of competence, flexibility, and efficiency, whereas interested patrons could not monitor or understand what was really going on without expert assistance. Again, mathematics served as an especially effective means not only to create new fields of expertise, by abstracting empirical practice into general principles, but also to establish the experts who controlled those fields as an obligatory point of passage. In the case of mathematical navigation, by the end of the sixteenth century neither patrons nor pilots were likely to have the level of mathematical learning needed to understand fully the new technologies being introduced, either where they came from or how they worked. Once established within English patronage networks, therefore, expert mediators could be very difficult to supplant.

The role of the expert mediator in facilitating centralized royal and corporate administration in England did not come to an end with Elizabeth's reign in 1603, of course. The seventeenth century saw privy councillors and wealthy investors take on projects of ever greater scale and expense throughout the realm, relying upon expert mediators more than ever before. One of the grandest of these undertakings was the drainage and reclamation of tens of thousands of marshy acres in England's Fenlands. Over several decades, wealthy investors (often including the king himself) sought to transform ever larger plots of common wetland across eastern England into dry farmland, hoping to amass huge fortunes in subsequent land speculation. The knowledge, skills, and experience required to plan and execute such projects were rare commodities in England, yet they were vital to those projects' success, creating a high demand for drainage experts among wealthy English patrons. In the course of this massive undertaking, drainage expertise became a powerful means of imposing royal governance and corporate control over remote areas of the realm that had long been comparatively resistant to them.[2] As they had throughout Elizabeth's reign, expert mediators continued to make their seventeenth-century patrons ever wealthier and more powerful, giving them greater control over valuable knowledge and resources and further consolidating the centralized governance and economic development of early modern England.

# Notes

INTRODUCTION: Expert Mediators and Elizabethan England

1. This overview of the Spanish Armada and the English response to it is based on Garrett Mattingly, *The Armada* (Boston: Houghton Mifflin, 1959); Colin Martin and Geoffrey Parker, *The Spanish Armada* (London: Hamilton, 1988); Duff Hart-Davis, *Armada* (London: Bantam, 1988); and Peter Padfield, *Armada: A Celebration of the Four Hundredth Anniversary of the Defeat of the Spanish Armada, 1588–1988* (London: Victor Gollancz, 1988). For an excellent annotated bibliography of the Armada, see Eugene L. Rasor, *The Spanish Armada of 1588: Historiography and Annotated Bibliography* (Westport, Conn.: Greenwood Press, 1993).

2. On the Spanish army in the Low Countries, see Geoffrey Parker, *The Army of Flanders and the Spanish Road, 1567–1659: The Logistics of Spanish Victory and Defeat in the Low Countries' Wars* (Cambridge: Cambridge University Press, 1972).

3. On sixteenth-century advances in English navigation, see David W. Waters, *The Art of Navigation in England in Elizabethan and Early Stuart Times* (London: Hollis and Carter, 1958). Thomas Cavendish's circumnavigation voyage, begun in 1586, was not actually completed until shortly after the Armada battle was over; he learned of the English victory on his way home.

4. John Knox Laughton, ed., *State Papers Relating to the Defeat of the Spanish Armada, Anno 1588* (1894; reprint, Brookfield, Vt.: Gower, 1987), 127–129.

5. Quoted in Mattingly, *Armada*, 216–217; Mattingly speculates that the source of the quotation may have been Juan Martínez de Recalde, one of the Armada's squadron commanders. Recalde managed to return to A Coruña harbor in Spain with his badly damaged ship, the *San Juan*, on 7 October 1588, but he died very shortly after his arrival.

6. Although many historians have written that fully half of the Armada was lost at sea, Garrett Mattingly has shown that this contemporary claim was probably based on faulty and incomplete information. *Armada*, 424–426.

7. On the development of English sailing and cannon-founding technologies in particular, see Carlo M. Cipolla, *Guns, Sails, and Empires: Technological Innovation and the Early Phases of European Expansion, 1400–1700* (New York: Minerva Press, 1965).

8. On the rise of the English Privy Council as a national, centralized administrative body, handling the day-to-day business of the central government while becoming ever more actively involved in local affairs throughout the sixteenth century, see G. R. Elton, *The Tudor Revolution in Government: Administrative Changes in the Reign of Henry VIII* (Cambridge: Cambridge University Press, 1953); Alan G. R. Smith, *The Emergence of a Nation State: The Commonwealth of England, 1529–1660* (London: Longman, 1984), chap. 13; Alan G. R. Smith, *The Government of Elizabethan England* (London: Edward Arnold, 1967); Joel Hurstfield, *Freedom, Corruption, and Government in Elizabethan England* (Cambridge, Mass.: Har-

vard University Press, 1973); Penry Williams, *The Tudor Regime* (Oxford: Clarendon Press, 1979); David Loades, *Power in Tudor England* (New York: St. Martin's Press, 1997), chap. 3; Michael Barraclough Pulman, *The Elizabethan Privy Council in the Fifteen-Seventies* (Berkeley and Los Angeles: University of California Press, 1971); A. Hassell Smith, *County and Court: Government and Politics in Norfolk, 1558–1603* (Oxford: Clarendon Press, 1974); and Diarmaid MacCulloch, *Suffolk and the Tudors: Politics and Religion in an English County, 1500–1600* (Oxford: Clarendon Press, 1986).

9. Hurstfield, *Freedom, Corruption, and Government*, 41–43; Lawrence Stone, *The Crisis of the Aristocracy, 1558–1641* (Oxford: Clarendon Press, 1965), 259–262, 708; see also Mark A. Kishlansky, *Parliamentary Selection: Social and Political Choice in Early Modern England* (Cambridge: Cambridge University Press, 1986). My own research on Dover in the 1580s provides a case in point; in 1583 the Privy Council ordered the removal and replacement of several town officials at Dover, including the mayor, because of their perceived opposition to central control of the harbor reconstruction. Privy Council to the jurats of Dover, 29 December 1583, Additional MSS, 29811, British Library, London, fols. 20–21.

10. Smith, *County and Court*, and MacCulloch, *Suffolk and the Tudors*, show how the Privy Council worked to gain influence in the context of local government by placing candidates loyal to them in key local offices, most especially those of the lords lieutenant and deputies of the local militia and the justices of the peace, and by attempting to diminish the power of traditional local officers, such as the sheriffs. Both, however, also demonstrate that the local authorities did not always yield voluntarily to the will of the privy councillors and their deputies and that the gentry could sometimes mount a successful resistance to the expansion of the council's local power base.

11. On technological patronage in western Europe, see Henry Heller, *Labour, Science and Technology in France, 1500–1620* (Cambridge: Cambridge University Press, 1996); Hélène Vérin, *La gloire des ingénieurs: L'intelligence technique du XVIe au XVIIIe siècle* (Paris: Albin Michel, 1993); David C. Goodman, *Power and Penury: Government, Technology, and Science in Philip II's Spain* (Cambridge: Cambridge University Press, 1988); Pamela H. Smith, *The Business of Alchemy: Science and Culture in the Holy Roman Empire* (Princeton, N.J.: Princeton University Press, 1994); Pamela O. Long, "Power, Patronage, and the Authorship of *Ars*: From Mechanical Know-How to Mechanical Knowledge in the Last Scribal Age," *Isis* 88 (1997): 1–41; Nicholas Adams, "Architecture for Fish: The Sienese Dam on the Bruna River—Structures and Designs, 1468–ca. 1530," *Technology and Culture* 25 (1984): 768–797; and J. R. Hale, "Tudor Fortifications: The Defence of the Realm, 1485–1558," in *The History of the King's Works*, ed. H. M. Colvin (London: H. M. Stationery Office, 1982), 4:367–401.

12. On the early modern history of joint-stock companies in Britain, see William Robert Scott, *The Constitution and Finance of English, Scottish, and Irish Joint-Stock Companies to 1720*, 3 vols. (1910–12; reprint, Gloucester, Mass.: Peter Smith, 1968).

13. Peter Dear, *Discipline and Experience: The Mathematical Way in the Scientific Revolution* (Chicago: University of Chicago Press, 1995), 76 n; Rose-Mary Sargent, "Scientific Experiment and Legal Expertise: The Way of Experience in Seventeenth-Century England," *Studies in the History and Philosophy of Science* 20 (1989): 28; *Oxford English Dictionary*, s.v. "expert." Dear points out that, according to the *Oxford English Dictionary*, *expert* was not used as a noun before the nineteenth century, but only as an adjective. One could be called an expert navigator, for example, but not an expert in the art of navigation. The word *expertise* was likewise coined during the nineteenth century.

14. "A narrative of the unwarrantable doings of Captain Frobisher," [1578?], Lansdowne MSS, 100/1, British Library, London, fol. 8 v.

15. The full quotation reads: "Wonderful many places, in the civil law, require an expert *Arithmetician*, to understand the deep judgment, and just determination of the ancient Roman lawmakers. But much more expert ought he to be, who should be able to decide with equity, the infinite variety of cases, which do, or may happen, under every one of those laws and ordinances civil." John Dee, "Mathematicall Praeface" to *The Elements of Geometrie of the most auncient Philosopher Euclide of Megara* . . . , trans. Henry Billingsley (London, 1570), fol. a.i v (original emphasis).

16. Letters patent of Queen Elizabeth, granting to Daniel Hechstetter and Thomas Thurland privilege to work royal mines, 10 October 1564, Sloane MSS, 1709, British Library, London, fol. 35 r.

17. Richard Eden, preface to Martín Cortés, *The Arte of Nauigation* . . . , trans. Richard Eden (London, 1561), fol. [C.iiii r].

18. On Tudor patronage strategies and the emphasis placed on practical action and results, see William H. Sherman, *John Dee: The Politics of Reading and Writing in the English Renaissance* (Amherst: University of Massachusetts Press, 1995); Michael A. R. Graves, *Thomas Norton, the Parliament Man* (Oxford: Blackwell, 1994); Lisa Jardine and Anthony Grafton, "'Studied for Action': How Gabriel Harvey Read His Livy," *Past and Present* 129 (1990): 30–78; Mordechai Feingold, *The Mathematicians' Apprenticeship: Science, Universities, and Society in England, 1560–1640* (Cambridge: Cambridge University Press, 1984), chap. 6; J. A. Bennett, "The Mechanics' Philosophy and the Mechanical Philosophy," *History of Science* 24 (1986): 1–28; J. A. Bennett, "The Challenge of Practical Mathematics," in *Science, Culture, and Popular Belief in Renaissance Europe*, ed. Stephen Pumfrey, Paolo L. Rossi, and Maurice Slawinski (Manchester: Manchester University Press, 1991); Stephen Johnston, "Mathematical Practitioners and Instruments in Elizabethan England," *Annals of Science* 48 (1991): 319–344.

19. On humanism in Tudor England, see James McConica, *English Humanists and Reformation Politics under Henry VIII and Edward VI* (Oxford: Clarendon Press, 1965); W. Gordon Zeeveld, *Foundations of Tudor Policy* (Cambridge, Mass.: Harvard University Press, 1948); Maria Dowling, *Humanism in the Age of Henry VIII* (London: Croom Helm, 1986); Alistair Fox and John Guy, *Reassessing the Henrician Age: Humanism, Politics, and Reform, 1500–1550* (New York: B. Blackwell, 1986); Winthrop Hudson, *The Cambridge Connection and the Elizabethan Settlement of 1559* (Durham, N.C.: Duke University Press, 1980); Thomas F. Mayer, *Thomas Starkey and the Commonweal: Humanist Politics and Religion in the Reign of Henry VIII* (Cambridge: Cambridge University Press, 1989); Feingold, *Mathematicians' Apprenticeship*.

20. On the humanist drive to abstract and codify practical knowledge, especially in the arts and sciences, see Michael Baxandall, *Painting and Experience in Fifteenth-Century Italy: A Primer in the Social History of Pictorial Style* (Oxford: Clarendon Press, 1972); Michael Baxandall, *Giotto and the Orators: Humanist Observers of Painting in Italy and the Discovery of Pictorial Composition, 1350–1450* (Oxford: Clarendon Press, 1971); Rudolph Wittkower, *Architectural Principles in the Age of Humanism*, (New York: St. Martin's Press, 1988); Paolo Rossi, *Philosophy, Technology, and the Arts in the Early Modern Era*, trans. Salvator Attanasio, ed. Benjamin Nelson (New York: Harper and Row, 1970); Anthony Grafton, *Leon Battista Alberti: Master Builder of the Italian Renaissance* (New York: Hill and Wang, 2000); David

Cast, "Humanism and Art," and Pamela O. Long, "Humanism and Science," in *Renaissance Humanism: Foundations, Forms, and Legacy*, ed. Albert Rabil Jr. (Philadelphia: University of Pennsylvania Press, 1988); Pamela O. Long, *Openness, Secrecy, Authorship: Technical Arts and the Culture of Knowledge from Antiquity to the Renaissance* (Baltimore: Johns Hopkins University Press, 2001); Pamela O. Long, "The Openness of Knowledge: An Ideal and Its Context in 16th-Century Writings on Mining and Metallurgy," *Technology and Culture* 32 (1991): 318–355; and Long, "Power."

21. Juan Luis Vives, *De tradendis disciplinis* (Antwerp, 1531); trans. and ed. Foster Watson, in *Vives: On Education: A Translation of the* De tradendis disciplinis *of Juan Luis Vives*, (1913; reprint, Totowa, N.J.: Rowman and Littlefield, 1971), 209.

22. Thomas More, *Utopia* (Louvain, 1516); trans. and ed. Robert M. Adams, 2d ed. (New York: W. W. Norton, 1992), 39.

23. François Rabelais, *La vie inestimable du grand Gargantua, père de Pantagruel* (c. 1534); trans. and ed. Burton Raffel, in *Gargantua and Pantagruel* (New York: W. W. Norton, 1990), 57, 62–63.

24. George Gascoigne, forward to *A Discovrse of Discouerie for a new Passage to Cataia*, by Humphrey Gilbert (London, 1576), fol. ¶¶ii r.

25. Humphrey Gilbert, "The erection of an Achademy in London for educacon of her Ma$^{tes}$ Wardes and others the youth of nobility and gentlemen," n.d., Lansdowne MSS, 98/1, fols. 2 r–7 r. Gilbert's manuscript treatise is undated, but it is endorsed by William Cecil after his elevation to the peerage as Baron Burghley in 1571. Gilbert himself was lost at sea on the return leg of a transatlantic voyage in 1583. The treatise has been edited by Frederick J. Furnivall and published as "Queene Elizabethes achademy . . ." (London: Early English Text Society, extra series 8, 1869; reprint, Millwood, N.Y.: Kraus Reprint, 1975), 1–12. All subsequent references are to the Lansdowne manuscript version, and those in the introduction are given parenthetically in the text.

26. Gilbert cited the instruction of the Cambridge Fellow John Cheke as a precedent for his proposal that students learn English oration for practical use. Although he was a renowned Greek scholar, Cheke eventually left Cambridge and became one of the Duke of Northumberland's principal clients at court during the reign of Edward VI. He was also instrumental in introducing a number of his former students to court, including his brother-in-law, William Cecil.

27. The Elizabethan Inns of Court, where England's professional lawyers obtained their training, admitted scores of gentlemen who meant to study the law as amateurs but the institutions did little or nothing to cater to their specific educational needs and limited legal interest (and often ability). See Wilfrid R. Prest, *The Inns of Court under Elizabeth I and the Early Stuarts, 1590–1640* (London: Longman, 1972), esp. 149–153.

28. Translation was a particular priority in Elizabethan England, for both patrons and clients. Gilbert, "The erection of an Achademy," fol. 6 r. There seems to have been a pervasive sense that Continental scholars possessed important knowledge that ought to be available to English readers in their vernacular. Translating a well-chosen work was one of the surest means of obtaining courtly literary patronage; see Eleanor Rosenberg, *Leicester, Patron of Letters* (New York: Columbia University Press, 1955), chap. 5.

29. For a few of the best-known examples, translated into English, see Leon Battista Alberti, *On painting and On sculpture: The Latin texts of* De pictura *and* De statua *[by] Leon Battista Alberti*, ed. and trans. Cecil Grayson (London: Phaidon, 1972); Leon Battista Alberti,

*On the Art of Building in Ten Books*, trans. Joseph Rykwert, with Neil Leach and Robert Tavernor (Cambridge, Mass.: MIT Press, 1988); Georgius Agricola, *De re metallica* (Basil, 1556); trans. Herbert Clark Hoover and Lou Henry Hoover (1912; reprint, New York: Dover Publications, 1950); *The Pirotechnia of Vannoccio Biringuccio*, trans. Cyril Stanley Smith and Martha Teach Gnudi (New York: Basic Books, 1959); and Agostino Ramelli, *The Various and Ingenious Machines of Agostino Ramelli (1588)*, trans. Martha Teach Gnudi (Baltimore: Johns Hopkins University Press, 1976).

30. Michael Baxandall has argued, for example, that Alberti composed his treatise *On Painting* not primarily for an audience of painters, but rather for an audience of humanist-educated patrons and critics. *Giotto and the Orators*, pt. 3.

31. See Long, *Openness*, chaps. 3–4; and Rossi, *Philosophy*, 32–33.

ONE: German Miners, English Mistrust, and the Importance of Being "Expert"

1. M. B. Donald, *Elizabethan Copper: The History of the Company of Mines Royal, 1568–1605* (London: Pergamon Press, 1955), 15–35.

2. Thurland to Cecil, 25 May 1566, State Papers, Domestic Series, Public Record Office, London (hereafter SPD), 12/39/80.

3. William Barclay Parsons, *Engineers and Engineering in the Renaissance* (Baltimore: Williams and Wilkins, 1939), 177–215.

4. Henry Hamilton, *The English Brass and Copper Industries to 1800*, 2d ed. (London: Cass, 1967), 2. In addition to copper, however, there are also lead veins in the area, which may have been what attracted the Romans' attention so far north.

5. C. N. Bromehead, "Mining and Quarrying to the Seventeenth Century," in *A History of Technology*, vol. 2, *The Mediterranean Civilizations and the Middle Ages, c. 700 BC to c. 1500 AD*, ed. Charles Singer, E. J. Holmyard, A. R. Hall, and Trevor I. Williams (Oxford: Clarendon Press, 1956), 10–12; John Nef, *The Conquest of the Material World* (Chicago: University of Chicago Press, 1964), 35–41. Copper was sporadically mined in England during the Middle Ages, but in comparison with the massive undertakings of central Europe, or even with contemporary tin, lead, and iron production in England, English copper mining was insignificant and technologically backward by 1560.

6. Parsons, *Engineers*, 179; Donald, *Elizabethan Copper*, 9–11. A few successful copper mines were also opened in Sweden after 1290.

7. Parsons, *Engineers*, 178–179.

8. Ibid., 179–181. The only major sixteenth-century work not written by a German was Vannoccio Biringuccio's *De la pirotechnia* . . . (Venice, 1540), which had five Italian and two French editions by the end of the seventeenth century. Biringuccio's work has been translated into English, in *Pirotechnia of Vannoccio Biringuccio*.

9. Georgius Agricola, *De re metallica* (Basil, 1556); trans. Herbert Clark Hoover and Lou Henry Hoover (1912; reprint, New York: Dover, 1950); all subsequent references are to the Hoover translation. See also Bern Dibner, *Agricola on Metals* (Norwalk, Conn.: Burndy Library, 1958), 26.

10. Herbert Clark Hoover, introduction to *De re metallica*, v–xvii; Parsons, *Engineers*, 179–181.

11. Agricola, *De re metallica*, 33–38.

12. George Hammersley, ed., *Daniel Hechstetter the Younger, Memorabilia and Letters, 1600–1639: Copper Works and Life in Cumbria*, (Stuttgart: Franz Steiner Verlag Wiesbaden GMBH, 1988), 99, 324–325, 336–337, and his introduction to the book, 64.

13. Various medieval grants of royal mining rights made specific mention of copper ore (e.g., Donald, *Elizabethan Copper*, 139–141), and it is telling that the original negotiations over mining rights for the Company of Mines Royal in 1561 covered almost all of the counties in which copper ore was subsequently found. Draft of the indenture between the queen, John Steynbergh, and Thomas Thurland, [16 July 1561], SPD, 12/18/18(II).

14. Parsons, *Engineers*, 178–179.

15. Agricola, *De re metallica*, 149–218; see 149 n. 1, for the ancient mechanical background.

16. Parsons, *Engineers*, 203–205, 212–213.

17. Ibid., 179, 213.

18. Ibid., 202.

19. Agricola, *De re metallica*, 219–544.

20. Parsons, *Engineers*, 179, 191–192. See also W. G. Collingwood, ed. and trans., *Elizabethan Keswick: Extracts from the Original Account Books, 1564–1577, of the German Miners, in the Archives of Augsburg*, tract series, no. 8 (Kendal, U.K.: Titus Wilson, 1912), for an overview of the organization of German mining administration in England, as well as extensive transcripts from the Germans' original account books, translated into English.

21. Agricola's book 1 included an account of the political, economic, and military benefits that mining brought to the realm and a description of various types of mining fraud; book 3 discussed the surveying and plotting of mine shafts above ground with a compass; book 4 explained how to determine the boundaries of mining rights, described the hierarchy and respective duties of various offices of mine management, and provided countless details of mine administration; book 5 discussed the principles of surveying underground passages by means of similar triangles.

22. Loner to William Burghley [Cecil], 1 March 1572, SPD, 12/85/63. William Cecil was created Baron Burghley on 25 February 1571. However, because most of the events this chapter describes took place before that date, to minimize confusion I refer to him as William Cecil throughout the text of the chapter.

23. Ibid.

24. "Memorial of the petition of Thomas Thurland and Sebastian Speydell to the queen," [1564], SPD, 12/34/59.

25. Hammersley, *Daniel Hechstetter*, 28–50, 62–64; Donald, *Elizabethan Copper*, 35–42; Collingwood, *Elizabethan Keswick*, 16–20, 120–126.

26. "Articles proposed for amendment of the charter of incorporation of the Company of Mines Royal," 10 September 1564, SPD, 12/34/58.

27. Ibid.

28. Ibid.

29. "Articles contained in the book of privilege to be granted to Thomas Thurland and Daniel Hechstetter," [1564], SPD, 12/34/60; see also the letters patent of Queen Elizabeth granting to Daniel Hechstetter and Thomas Thurland privilege to work royal mines, 10 October 1564, Sloane MSS, fols. 35–38. The agreement stipulated that the queen was to receive one-tenth of all gold and silver, but the royal share of copper was variable.

30. Donald, *Elizabethan Copper*, 103.

31. Thurland and Hechstetter to Cecil, 26 May 1565, SPD, 12/36/59. Although the Privy Council was not officially connected with the Company of Mines Royal, William Cecil, the council's principal secretary, was the addressee of the vast majority of the company's correspondence, and virtually all of the surviving documents relating to the company are now included in the various archives of Elizabethan state papers that he kept.

32. John Roche Dasent, ed., *Acts of the Privy Council of England*, n.s. (London: H. M. Stationery Office, 1890–1949), 7:229 (8 July 1565). About 150 German miners and managers emigrated to England during the sixteenth century, many of them becoming permanent denizens.

33. Humfrey was not appointed assay master until 14 July 1561, after the refining of the English coinage was well under way. His predecessor in the office, John Bull, was a London wool merchant who vacated the post when he was promoted to the office of controller of the mint; see M. B. Donald, *Elizabethan Monopolies: The History of the Company of Mineral and Battery Works from 1565 to 1604* (Edinburgh: Oliver and Boyd, 1961), 24.

34. Humfrey to Cecil, 12 May 1565, SPD, 12/36/49. The hundredweight was a variable measure of weight in England with a long and complex history, but for the purpose of measuring copper in the sixteenth century, it was equivalent to 112 pounds. "Certificate of two justices of Cumberland, of having received of the miners for the queen's use five hundred quintels of copper," 23 December 1574, Lansdowne MSS, 18/53. See R. D. Connor, *The Weights and Measures of England* (London: Science Museum, 1987), 135–138.

35. Thurland to Cecil, 20 March 1565, SPD, 12/36/25.

36. Hechstetter to Earl of Pembroke, Earl of Leicester, and Cecil, 29 September 1567, SPD, 12/44/16. Continental ores are composed mostly of copper oxide and carbonate and are comparatively easy to smelt. The ores found in England are composed primarily of copper sulfide, however, which requires considerable prior preparation and a much higher temperature for smelting correctly. See R. J. Forbes, "Metallurgy," in *A History of Technology*, vol. 2, *The Mediterranean Civilizations and the Middle Ages, c. 700 BC to c. 1500 AD*, ed. Charles Singer, E. J. Holmyard, A. R. Hall, and Trevor I. Williams (Oxford: Clarendon Press, 1956), 48–50.

37. Thurland to Cecil, 20 March 1565, SPD, 12/36/25.

38. Humfrey to Cecil, 12 May 1565, SPD, 12/36/49.

39. On the early modern market for finished copper products, see G. Hammersley, "Technique or Economy? The Rise and Decline of the Early English Copper Industry, ca. 1550–1660," in *Schwerpunkte der Kupferproduktion und des Kupferhandels in Europa, 1500–1650*, ed. Hermann Kellenbenz (Cologne: Böhlau Verlag GmbH, 1977), 27–35; Hamilton, *English Brass*, 25–27.

40. Humfrey to Cecil, 27 November 1565, SPD, 12/37/73.

41. Hechstetter to Earl of Pembroke, Earl of Leicester, and Cecil, 29 September 1567, SPD, 12/44/16.

42. George Nedham to Cecil, 25 May 1567, SPD, 12/42/68; Daniel Ulstatt to Cecil, 25 June 1568, SPD, 12/46/80.

43. Humfrey to Cecil, 12 May 1565, SPD, 12/36/49.

44. Humfrey to Cecil, 23 May, 2 July, 27 November 1565, SPD, 12/36/58, 12/36/73, 12/37/73.

45. Humfrey to Cecil, 2, 18 July; 16, 28 August 1565, SPD, 12/36/73, 12/36/82, 12/37/5, 12/37/21.

46. Humfrey to Cecil, 2 July, 28 August, 15 September 1565, SPD, 12/36/73, 12/37/21, 12/37/30 (quotations in the letters of 2 July and 28 August).

47. Humfrey to Cecil, 27 November 1565, SPD, 12/37/73; see also Humfrey to Cecil, 2 July, 28 August 1565, SPD, 12/36/73, 12/37/21.

48. For more information on the Company of Mineral and Battery Works, see Scott, *Constitution and Finance*, 2:413–429; and Donald, *Elizabethan Monopolies*.

49. W. R. Scott, "The Constitution and Finance of an English Copper Mining Company in the Sixteenth and Seventeenth Centuries: Being an Account of 'The Society of the Mines Royal,'" *Vierteljahrschrift für Social und Wirtschaftsgeschichte* 5 (1907): 525–552.

50. Hechstetter and Loner to Lionel Duckett, 23 April 1566, SPD, 12/39/57. William Cecil, in contrast, had paid the calls on his two shares promptly.

51. Hechstetter and Loner to Duckett, 23 April 1566, SPD, 12/39/57.

52. Hechstetter and Loner to Cecil, 25 June 1566, SPD, 12/40/14.

53. Ibid.

54. Hechstetter to Loner, 10 October 1566, SPD, 12/40/81. The report in the State Papers collection is labeled as a copy; it is not known how the English partners obtained a copy of a letter sent from Hechstetter to Loner, or whether Hechstetter composed his original letter in English or German.

55. Ibid. W. G. Collingwood relates a very different version of this story; he points out that anyone buried at Keswick was supposed to be recorded in the parish register of Crossthwaite parish. That register contains an entry for the burial in September 1566 of "Leonard and John Stilte, infants." Collingwood surmises that since one Jobst Stoltz is later referred to in the same register as "Stilt," and since there is no other mention of a Leonard with a German surname, the Leonard Stoultz in question must have been a baby. He describes a hypothetical scenario in which some English tempers flared, and a chance rock was thrown which unfortunately struck and killed a very young child: "the accidental result of some casual rough behavior, not an organized attack on the miners," and certainly not an intentional murder. Hechstetter's report, he suggests, was therefore "greatly exaggerated through ... natural anxiety," especially considering that Hechstetter was not present in Keswick during the incident. Collingwood, *Elizabethan Keswick*, 17–18. It should be noted that although relations between the German miners and English locals certainly did erupt in occasional violence, for the most part the English attitude toward the strangers was surprisingly peaceful, especially for such a remote corner of the realm, unaccustomed to receiving foreign visitors. Roughly one hundred of the Germans chose to remain in England all their lives. By the end of 1567, there had been sixteen marriages between German miners and English women, and by 1600 there had been at least sixty. Moreover, Daniel Hechstetter's children married into some of the most prominent families in northern England. See Collingwood, *Elizabethan Keswick*, 16–20; Hammersley, *Daniel Hechstetter*, 62–64.

With respect to the Stoultz incident, we will probably never know precisely what happened. I am not as ready as Collingwood to dismiss Hechstetter's version of events as being "greatly exaggerated." The reference to "Leonard Stilte" as an "infant" in the Crossthwaite parish register does not necessarily imply that he was a small child; according to the *Oxford English Dictionary*, s.v. "infant," *infant* could refer to anyone who was under the legal age of twenty-one. According to this definition, the individual in question could well have been a young man working in the mines, who was set upon and murdered by a gang of English ruffians, much as Hechstetter described. Whatever took place, it was certainly regarded by most of the English as a tragedy. Lady Radcliffe's role in the incident notwithstanding, a number of the English authorities were very upset; the bishop of Carlisle, for one, was in favor of executing the offenders. Hechstetter to Loner, 10 October 1566, SPD, 12/40/81. Lionel Duckett wrote a letter to Cecil asking for help in suppressing unrest and punishing those who had attacked the Germans. Duckett to Cecil, 13 October 1566, SPD, 12/40/83. And the queen herself issued a very stern warning to the local justices of the peace, stating that the Germans were in England at her invitation and under her protection and that they were not to be further molested in any way, on pain of royal wrath. Queen to Lord Scrope and the justices of the peace in Westmoreland and Cumberland, 16 October 1566, SPD, 12/40/87.

56. Hechstetter to Loner, 10 October 1566, SPD, 12/40/81.

57. Ibid.

58. On "economic nationalism" in early modern England, see Liah Greenfeld, *The Spirit of Capitalism: Nationalism and Economic Growth* (Cambridge, Mass.: Harvard University Press, 2001), 29–58, 112–114.

59. Thurland to Cecil, 25 May 1566, SPD, 12/39/80. The original source reads "Doutche" and "Doutchman"; *Dutch* was a common sixteenth-century English term for *German*, an anglicized form of *Deutsch*.

60. Donald, *Elizabethan Copper,* 83–84. A brief contemporary history of the search for gold in Scotland through the sixteenth century, and an assessment of the likelihood of mining it at a profit, may be found in the Cotton MSS, British Library, London, Otho E/X/12.

61. Thurland to Cecil, 7 October 1566, SPD, 12/40/79; Thurland to the queen, 7 October 1566, SPD, 12/40/80.

62. Thurland to Cecil, 7 October 1566, SPD, 12/40/79; Thurland to the queen, 7 October 1566, SPD, 12/40/80.

63. Thurland's letter to the queen (7 October 1566, SPD, 12/40/80) clearly portrays De Vos as behaving in a secretive and underhanded way; his letter to Cecil of the same date (SPD, 12/40/79), however, is more ambiguous. As in his letter to the queen, Thurland informed Cecil that De Vos "made proof [of the soil sample] secretly with my strangers, which he made me not privy to." Yet he also wrote to Cecil that De Vos was somewhat more forthcoming with vague information: "Notwithstanding he told me if that [sample] were good he could get before Xmas next ten [thousand] marks worth of it and willed me [to] keep it secret [until] his return." This version of the story would seem to imply that Thurland learned about the possibility of gold at Crawford Moor from De Vos, but not the outcome of the German's assay—this could help to explain why he felt the need to steal some of De Vos's sand and have it assayed for himself. Regardless, Thurland was consistent in claiming that De Vos was withholding key information from him, and the purpose of his letters was to keep the queen and the company's shareholders as well informed as he could: "If he [De Vos] make you not privy, send me word [and] I will come post to prevent all their doings." SPD, 12/40/79.

64. Thurland to the queen, 7 October 1566, SPD, 12/40/80.

65. Thurland to Cecil, 5 December 1566, SPD, 12/41/38.

66. Ibid.

67. Nedham to Duckett, 13 October 1568, SPD, 12/48/14.

68. Donald, *Elizabethan Copper,* 83–89.

69. For a detailed discussion of *Queen v. Northumberland,* see Eric H. Ash, "Queen v.

Northumberland, and the Control of Technical Expertise," *History of Science* 39 (2001): 215–240.

70. Hechstetter to Cecil, 29 April 1570, summarized in the *Calendar of the Manuscripts of the Most Hon. the Marquess of Salisbury . . . preserved at Hatfield House, Hertfordshire* (London: H. M. Stationery Office, 1883), 1:467, no. 1484; Duckett to Burghley [Cecil], 29 March 1571, SPD, 12/77/37; Hamilton, *English Brass*, 25–27; Scott, "English Copper Mining," 535.

71. Humfrey to Cecil, 20 December 1566, SPD, 12/41/45; Hamilton, *English Brass*, 25–26.

72. The primary and secondary accounts of the company's income and expenses during the sixteenth century are numerous, confusing, and often contradictory, making it difficult to determine the precise details of the company's financial situation. The overall trend, however, remains consistent: throughout the 1570s, the company produced copious amounts of copper, but the market value to be had for it could not cover the very high costs of its production, so that by 1580 the company was in a severe financial crisis. See Donald, *Elizabethan Copper*, 216–258; Hammersley, "Technique," 4–5; Hamilton, *English Brass*, 20–27; and Scott, "English Copper Mining," 530–541.

73. Loner to Cecil, 27 October 1570, SPD, 12/74/21.

74. Hans Langnauer to Loner, 22 September 1570, translated and quoted in Collingwood, *Elizabethan Keswick*, 97.

75. Loner to Cecil, 27 October, 13 December 1570, SPD, 12/74/21, 12/74/47.

76. Langnauer to Loner, 22 September 1570, translated and quoted in Collingwood, *Elizabethan Keswick*, 97.

77. Loner to Cecil, 27 October 1570, SPD, 12/74/21; the right margin of the first page of Loner's letter is damaged, and some of the text is missing.

78. Nedham to Cecil, 30 November 1570, SPD, 12/74/43; Duckett to Burghley [Cecil], 7 March 1571, SPD, 12/77/29.

79. Unsigned letter to Cecil, n.d., SPD, 12/74/56. The authorship of this letter is uncertain. The *Calendar of State Papers, Domestic Series*, speculates that it was written by Daniel Hechstetter, but this seems unlikely because the author consistently refers to the Germans in the third person and the English in the first, which would seem to indicate an English author. I believe that the author was probably Thomas Thurland, though it is not in his normal hand; it could well be a copy of the original, a means by which several of Thurland's letters were preserved—this could also explain the missing signature.

80. Summary of Ulstatt's letter from Keswick, 1 October 1568, SPD, 12/48/1; unsigned letter to Cecil, n.d., SPD, 12/74/56 (quotation from the unsigned letter).

81. Unsigned letter to Cecil, n.d., SPD, 12/74/56.

82. Loner to Burghley [Cecil], 1 March 1572, SPD, 12/85/63.

83. Scott, "English Copper Mining," 534; Hamilton, *English Brass*, 28.

84. Scott, "English Copper Mining," 540–541.

85. Humfrey to Cecil, 22 July 1565, SPD, 12/36/83.

TWO: Expert Mediation and the Rebuilding of Dover Harbor

1. Reginald Scot, "The note of Reginald Scot esquier concerning Dover haven," in *Holinshed's Chronicles of England, Scotland, and Ireland . . .* , ed. John Hooker (1586; reprint,

London: J. Johnson et al., 1808), 4:845–868. Scot's chronicle is an excellent contemporary account of the project and gives details that could only have come from someone personally familiar with it, nicely supplementing the records in the Elizabethan State Papers. It should be noted, however, that Scot himself had helped to carry out some of the project's more controversial decisions and was the cousin of one of the works' chief overseers, so he was not an unbiased observer. For additional subsequent praise of the new harbor, see Lord Admiral Charles Howard of Effingham to William Burghley, 2 November 1593, SPD, 12/246/1.

2. For a thorough description of Dover Harbor, its location and the problems deriving therefrom, and reproductions of several contemporary plans of the harbor, see William Minet, "Some Unpublished Plans of Dover Harbour," *Archaeologia* 72 (1922): 185–224; much of this section of the chapter is adapted from Minet's account. See also Martin Biddle and John Summerson, "Dover Harbour," in *The History of the King's Works*, ed. H. M. Colvin, John Summerson, Martin Biddle, J. R. Hale, and Marcus Merriman (London: H. M. Stationery Office, 1982), 4:729–768; and Stephen Johnston, "Making Mathematical Practice: Gentlemen, Practitioners, and Artisans in Elizabethan England" (D.Ph. diss., Cambridge University, 1994), chap. 5.

3. Minet, "Some Unpublished Plans," 188; Minet points out that Julius Caesar's description of an attempted landing site during his first invasion of Britain corresponds very closely to the geography at Dover and that the Roman lighthouse that still stands within the castle grounds is the only surviving example of such a structure.

4. Camber, Winchelsea, and Rye (all located on the same estuary, southwest of Dover), though once thriving commercial ports, are now separated from the coast by roughly a mile of open fields, the cumulative result of the tidal silting that is so prevalent in the Channel. This phenomenon was not limited to the English coastline; on the gradual destruction of Harfleur Harbor and the subsequent adjacent construction of Le Havre in Normandy during the sixteenth century, see Parsons, *Engineers*, 460–477.

5. John Lyon, *The History of the Town and Port of Dover, and of Dover Castle; with a short account of the Cinque Ports* (Dover, U.K.: Ledger and Shaw; London: Longman, Hurst, Rees, Orme, and Browne, 1813), 1:150; Scot, "Dover haven," 846. Biddle and Summerson contend, however, that there is little or no evidence that the king ever sent money or otherwise supported Clark's works. "Dover Harbour," 730.

6. Mayor and jurats of Dover to Thomas Cromwell, 26 November 1533, in *Letters and Papers, Foreign and Domestic, of the Reign of Henry VIII*, ed. James Gairdner (London: H. M. Stationery Office, 1888), vol. 6, no. 1472.

7. Scot, "Dover haven," 847; Biddle and Summerson, "Dover Harbour," 732. Scot wrote that the petition was sent in 1532, although the first mention of Thomson and his petition in the *Letters and Papers* is on 26 November 1533 (vol. 6, no. 1472).

8. Minet, "Some Unpublished Plans," 192–193. Reginald Scot claimed that fifty thousand pounds was spent on the works ("Dover haven," 848), and William Lambarde, the pioneering historian of Kent, put the figure as high as sixty-three thousand (William Lambarde, *A Perambulation of Kent* . . . [1576; reprint, Chatham: W. Burrill; London: Baldwin, Cradock, and Joy, 1826], 133–134), whereas Biddle and Summerson place the cost at "at least £51,045" ("Dover Harbour," 736). These were enormous sums of money; in comparison, the entire Elizabethan project, which took place fifty years later (in a century of high inflation) and was far more ambitious, cost less than one-third as much.

9. William Borough, "Notes touching Dover haven," [1580], Lansdowne MSS, 22/10; Johnston, "Making Mathematical Practice," 227–228. The surveyors were also commissioned in 1576 to carry out similar investigations at Rye, Winchelsea, and Camber harbors. Borough reported that the latter two were beyond repair (in 1558 Camber accommodated fifty-two ships at low tide; in 1576 it was dry) but that Rye might still be saved if quick action was taken. There is no evidence, however, that the Elizabethan Privy Council ever attempted to salvage the harbor at Rye.

10. Dasent, *Acts*, 11:54–55 (21 February 1579).
11. Ibid., 11:55 (21 February 1579); 12:161 (15 August 1580).
12. Scot, "Dover haven," 850.
13. Biddle and Summerson, "Dover Harbour," 757–758. On the sixteenth-century Privy Council's tendency to rely upon trusted members of the local gentry for management of local affairs, see Smith, *County and Court*; and MacCulloch, *Suffolk and the Tudors*.
14. Leonard Digges and Thomas Digges, *A Geometrical Practise, named Pantometria* . . . (London, 1571); Thomas Digges, *Alae seu Scalae Mathematicae* (London, 1573); Thomas Digges, *A Perfit Description of the Caellestiall Orbes*, published with Leonard Digges and Thomas Digges, *A Prognostication Euerlasting* (London, 1576); and Leonard Digges and Thomas Digges, *An Arithmeticall Militare Treatise, named Stratioticos* . . . (London, 1579). For biographical information on Thomas Digges (d. 1595), see Francis R. Johnson, *Astronomical Thought in Renaissance England: A Study of the English Scientific Writings from 1500 to 1645* (Baltimore: Johns Hopkins Press, 1937), esp. chap. 6; Johnson and Sanford V. Larkey, "Thomas Digges, the Copernican System, and the Idea of the Infinity of the Universe in 1576," *Huntington Library Bulletin* 5 (1934): 69–117; E. G. R. Taylor, *Tudor Geography, 1485–1583* (London: Methuen, 1930); E. G. R. Taylor, *The Mathematical Practitioners of Tudor and Stuart England* (Cambridge: Published for the Institute of Navigation at the University Press, 1954), 175; Johnston, "Making Mathematical Practice," esp. chap. 2.
15. Mayor and jurats of Dover to the Privy Council, n.d., SPD, 12/120/24. On early modern Dutch hydraulic construction, see Petra J. E. M. van Dam, "Ecological Challenges, Technological Innovations: The Modernization of Sluice Building in Holland, 1300–1600," *Technology and Culture* 43 (2002): 500–520.
16. Dover commission to Privy Council, 18 August 1579, SPD, 12/131/72; Borough, "Notes touching Dover haven," [1580], Lansdowne MSS, 22/10; Johnston, "Making Mathematical Practice," 228.
17. Dover commission to Privy Council, 18 August 1579, SPD, 12/131/72; Johnston, "Making Mathematical Practice," 228.
18. Philip Chilwell de la Garde, "On the Antiquity and Invention of the Lock Canal of Exeter," *Archaeologia* 28 (1840): 7–26. De la Garde wrote that Trew's canal apparently accommodated vessels of up to sixteen tons, which was more than Trew had promised. The city's governors were nevertheless unhappy with his work, perhaps because the canal was not navigable at all tides, and they sued him. Trew was involved in litigation over his role in building the Exeter canal for many years. See also Johnston, "Making Mathematical Practice," 277–278.
19. Borough, "Notes touching Dover haven," [1580], Lansdowne MSS, 22/10; Scot, "Dover haven," 851.
20. Scot, "Dover haven," 851. Although Trew planned to incorporate the foundation laid for Thomson's pier during Henry VIII's reign into his own works, such a heavy wall

would most likely have settled into the deep mud and sand of Dover Bay, eventually destroying the wall's structural integrity.

21. Dasent, *Acts*, 12:161 (15 August 1580). When the Privy Council consulted Borough in 1580 regarding Trew's plan for repairs, he did not reject it out of hand. Believing that Trew's proposed wall might yield some benefit, even "though it work not the effect that he promised," Borough initially gave the plan his half-hearted support, "the rather for the easiness of the charge, and to keep life in that action, which otherwise I doubted would die, and Dover be left without hope of relief." However, he had no illusions that Trew's works would provide a full and permanent solution to the silting problem, and he ultimately condemned the plan, calling it "frivolous and vain." Borough, "Notes touching Dover haven," [1580], Lansdowne MSS, 22/10.

22. Dover commission to Privy Council, 24 August 1580, SPD, 12/141/36. On premodern English weights and measures, see Ronald Edward Zupko, *British Weights and Measures: A History from Antiquity to the Seventeenth Century* (Madison: University of Wisconsin Press, 1977), 21.

23. Cobham to Walsingham, 1 September 1580, SPD, 12/142/1; "Sundry points for the consideration of the council," 1 September 1580, SPD, 12/142/2; Requests of Mr. Trew, 1 September 1580, SPD, 12/142/3. As principal secretary, Walsingham was the main point of contact between the Privy Council and the Dover commission, the man to whom virtually all Dover-related correspondence was addressed.

24. Dasent, *Acts*, 12:197 (11 September 1580). The terms of Trew's commission are preserved in the queen's letter notifying all justices of the peace, 11 September 1580, SPD, 12/142/16.

25. Borough, "Notes touching Dover haven," [1580], Lansdowne MSS, 22/10; [Thomas Digges], "Notes touching the decay and remedy of Dover haven," [1580?], Lansdowne MSS, 22/11; Thomas Digges, "The Commission's reasons for condemning the proceedings of John Trew in the works of Dover haven," [1581], Lansdowne MSS, 31/74; Richard Barrey to Walsingham, 6 January 1582, SPD, 12/152/2; Trew quoted in Scot, "Dover haven," 851; Johnston, "Making Mathematical Practice," 275–286.

26. Dasent, *Acts*, 13:80 (14 June 1581), 13:139–140 (26 July 1581); quotation in "Notes of what is thought meet to be proceeded unto by the Dover commissioners," March 1583, SPD, 12/159/52; Johnston, "Making Mathematical Practice," 276. Unfortunately, the original Acts of the Privy Council for 1582 though 1585 are lost, so no record is available from the council's perspective of the works undertaken during that period.

27. Dasent, *Acts*, 12:161 (15 August 1580).

28. Scot, "Dover haven," 851.

29. Thomas Digges, "Brief notes of his proceedings in the works at Dover harbor and the present state thereof," 8 June 1584, SPD, 12/171/13(I). Digges's formal plan survives as "A briefe discourse declaringe how honorable and profitable to youre most excelle[n]t maiestie, and howe necessary and com[m]odiouse for your realme, the making of Douer Haven shalbe, and in what sorte, w[ith] leaste charge in greateste perfection, the same maye be accomplyshed," in the archives of the Society of Antiquaries of London; it was edited by T. W. Wrighte and published in *Archaeologia* 11 (1794): 212–254. Copies of some parts of the plan may also be found scattered through the State Papers.

30. Scot, "Dover haven," 851–855.

31. Dover commission to Privy Council, 21 April 1582, SPD, 12/153/15; "Articles

agreed upon between the Privy Council and Fernando Poyntz," 27 June 1582, SPD, 12/154/20; Biddle and Summerson, "Dover Harbour," 757.

32. Privy Council to Dover commission, 14 April 1582, SPD, 12/153/8; "Articles agreed upon between the Privy Council and Fernando Poyntz," 27 June 1582, SPD, 12/154/20.

33. Bedwell's name first appears in the State Papers in April 1582 (SPD, 12/153/27), when he offered his Dover plans to the Privy Council, as well as proposals for various mathematical measuring instruments, and even a water clock to be used in determining longitude at sea. While at Dover he took every opportunity to complain that Poyntz was incompetent and never took his (Bedwell's) advice (see, for example, Barrey to Walsingham, 12 December 1582, SPD, 12/156/14; Bedwell to Burghley and Walsingham, 19 December 1582, SPD, 12/156/23–24). For more on Bedwell, see Johnston, "Mathematical Practitioners," 319–344.

34. Digges was especially close to the lord treasurer, William Burghley, to whom he dedicated his astronomical treatise *Alae seu Scalae Mathematicae* in 1573. Digges even credited Burghley with inspiring his systematic astronomical investigation of the 1572 supernova when the latter asked him to explain the astrological significance of the new star. Digges to Burghley, 11 December 1572, SPD, 12/90/12. After 1573 Digges remained a favored client of Burghley's in both parliamentary and private contexts; see Johnston, "Making Mathematical Practice," 60–62.

35. "Comparative estimate of the charges for the repair of Dover Haven," 15 May 1582, SPD, 12/153/51; "Certain articles to be considered by the Commissioners for Dover Haven," [January 1583?], SPD, 12/158/47. In defense of Poyntz's honesty, it appears that the mistake was a genuine misunderstanding; within a few weeks of being asked to do so, he had made a full account of his expenditures to the Privy Council's satisfaction and was allowed to continue with the works through the summer.

36. Bedwell to Burghley and Walsingham, 19 December 1582, SPD, 12/156/23–24; Barrey to Privy Council, 29 December 1582, SPD, 12/156/33.

37. Barrey to Privy Council, 29 December 1582, SPD, 12/156/33; Poyntz to Walsingham, 2 February 1583, SPD, 12/158/53.

38. Poyntz to Walsingham, 19 December 1582, 2 February 1583, SPD, 12/156/22, 12/158/53; Scott to Walsingham, 18 May 1582, SPD, 12/153/53; Mayor and jurats of Dover to Privy Council, 3 February 1583, SPD, 12/158/54.

39. Barrey to Walsingham, 12 December 1582, SPD, 12/156/14; Barrey to Cobham, 26 January 1583, SPD, 12/158/25; Report by William Borough on Poyntz's works, 10 February 1583, SPD, 12/158/61; Dover commission to Privy Council, 8 March 1583, SPD, 12/159/11(I).

40. Plan of the completed harbor at Dover, 1595, Cotton MSS, Augustus I/i/46 (no. 11).

41. Scot, "Dover haven," 853; "Report and suggestions by William Borough," 18 February 1583, SPD, 12/158/72.

42. The kinsman in question was almost certainly Reginald Scot, the chronicler of the Dover project, who had served as surveyor of the Romney seawall construction works and was also Thomas Scott's cousin.

43. Scot, "Dover haven," 860; see also Biddle and Summerson, "Dover Harbour," 759.

44. Digges, "A briefe discourse," 222–225.

45. [Digges], "Notes touching the decay and remedy of Dover haven," [1580?], Lansdowne MSS, 22/11. Although this document is unsigned, Thomas Digges is almost certainly the author. The handwriting is very similar to a 1582 report of Digges's on the Dutch plans to rebuild the harbor (SPD, 12/152/87), and the author makes several arguments here that are echoed in Digges's other proposals. Interestingly, the author of this plan cited for his model not the seawalls at Romney but rather various harbors throughout northern Europe, including "Ostende in Flanders" and "Newhaven, Flushing, and many other harbors of the Low Countries." Digges later wrote to the Privy Council that his travels in the Low Countries had originally inspired his own plan for Dover harbor: "[B]y precedent of the Low Countries the pent might be made with earth bays, of such substances as were to be found there [at Dover]." Digges, "Brief notes," 8 June 1584, SPD, 12/171/13(I).

46. "Answers to the considerations for placing and making the long wall at Dover haven," March 1582, SPD, 12/152/86.

47. "Answer of the Commissioners for Dover Haven to the six articles sent down by the Council," 8 March 1583, SPD, 12/159/12.

48. Ibid.

49. Boys to Walsingham, 25 March 1583, SPD, 12/159/45; Digges to Walsingham, 20 March 1583, SPD, 12/159/26; Scott to Walsingham, 24 March 1583, SPD, 12/159/44.

50. Scott to Walsingham, 24 March 1583, SPD, 12/159/44.

51. "Articles concerning the walls for the pent of Dover Harbour," 29 March 1583, SPD, 12/159/49; "The answers of the men of Romney Marsh," 29 March 1583, SPD, 12/159/50.

52. From at least 1572 onward, and throughout his tenure as master surveyor, Thomas Digges was politically very active in London. He worked closely with the Privy Council as a member of Parliament, especially in addressing the thorny issue of the Elizabethan succession crisis. The problem in that case was how best to protect the queen from her Catholic enemies and the conspiring partisans of Mary, Queen of Scots, the Catholic heir apparent to the English throne. In 1584–85 Digges played a central role in formulating the government's planned response to any potential assassination attempt upon Elizabeth, helping to turn the vigilante Oath of Association into a more formal and legal arrangement, the 1585 "Act for the Surety of the Queen's Most Royal Person." Thomas Digges, "Consideration of the dangers that may ensue by the Oath of Allegiance," [1585], SPD, 12/176/26; another copy of the same, January 1585, Lansdowne MSS, 98/4; Thomas Digges, "A brief discourse to the queen against succession known," 1585, SPD, 12/176/32. Among other things, Digges proposed (with the Privy Council's approval) that Parliament should never be dissolved during Elizabeth's lifetime and that the members should act as a provisional government upon her death, deciding the succession if she had no legitimate heirs. This affair has been seen as a watershed in Tudor government and republican political philosophy; see J. E. Neale, *Elizabeth I and Her Parliaments, 1584–1601* (London: Jonathan Cape, 1957), 28–57; Patrick Collinson, "The Monarchical Republic of Elizabeth I," *Bulletin of the John Rylands University Library of Manchester* 69 (1986): 394–424; and Patrick Collinson, "The Elizabethan Exclusion Crisis and the Elizabethan Polity," *Proceedings of the British Academy* 84 (1993): 51–92.

53. Digges, "Brief notes," 8 June 1584, SPD, 12/171/13(I).

54. "Digges's report of his proceedings in the works at Dover Haven, from their commencement," n.d., SPD, 12/175/18.

55. Scot, "Dover haven," 851.
56. "Digges's report of his proceedings in the works at Dover Haven, from their commencement," n.d., SPD, 12/175/18.
57. Digges, "Brief notes," 8 June 1584, SPD, 12/171/13(I); Digges did not say who had offered him the money.
58. Scot, "Dover haven," 854–855.
59. Digges, "Brief notes," 8 June 1584, SPD, 12/171/13(I). This document actually contains only Digges's adversarial characterization of the wood advocates' position. However, a new plan and estimate (of £14,420) from Pett and Baker, using wooden walls and submitted with Borough's approval, offered explicit criticism of the Romney method. "Estimate by William Borough of the charge for the works at Dover Haven," 3 April 1583, SPD, 12/160/3.
60. Scot, "Dover haven," 855.
61. Poyntz to Walsingham, 25 April 1583, SPD, 12/160/22; Scot, "Dover haven," 856. As it happened, Poyntz's assigned portion of the seawall was not completed until the early 1590s, being of low priority.
62. Motions made to the Privy Council, 10 April 1583, SPD, 12/160/6(I).
63. Scott to Walsingham, 31 May, 2 June 1583, SPD, 12/160/61, 12/161/2.
64. Scot, "Dover haven," 861.
65. Ibid., 865. Scot claimed that fewer than four days in all were lost to foul weather, unusual in England to say the least, and also related tales of miracles, including one "I myself have seen" of a fully laden cart "pass[ing] over the belly or stomach of the driver, and yet he not hurt at all thereby."
66. Dover commission to Walsingham, 7 June 1583, SPD, 12/161/4; Scott and Barrey to Walsingham, 9 June 1583, SPD, 12/161/7; Pett and Baker to Walsingham, 9 June 1583, SPD, 12/161/8.
67. Scott to Walsingham, 21 July 1583, SPD, 12/161/39.
68. Boys and Palmer to Walsingham, 8 December 1583, SPD, 12/164/13.
69. "Articles to be considered and resolved on touching Dover Haven," n.d., SPD, 12/170/98.
70. "Reasons to prove the East mouth of the haven at Dover to be more commodious than the West," n.d., SPD, 12/170/97.
71. "Questions demanded of the Masters and Mariners sent for by the Mayor of Dover," n.d., SPD, 12/170/99.
72. Ibid.
73. Collection of documents sent from Digges to Walsingham, 8 June 1584, SPD, 12/171/13(I–V).
74. "Resolution and opinions of the Commissioners," 16 June 1584, SPD, 12/171/30(I). Henry Palmer did not record any vote; presumably he did not attended the meeting.
75. Hawkins to Walsingham, 17 June 1584, SPD, 12/171/33; Hawkins, Borough, and Pett to Walsingham, 19 June 1584, SPD, 12/171/41.
76. Kishlansky, *Parliamentary Selection;* see especially chaps. 3–4 on the preference for consensus over polling among local gentry when filling a contested parliamentary seat.
77. John Tooke, "A history of Dover harbor," 1604, Additional MSS, 12514, fol. 39 r. This history, written by a former mayor and jurat of Dover, purportedly covers the entire period between Julius Caesar's landing and the reign of Elizabeth but focuses especially on

the rebuilding effort of the 1580s. In his lengthy account, Tooke claims to have been personally involved in planning and overseeing the works; he includes a detailed account of the harbor mouth debate, in which he allegedly played a prominent role. His report is spotty and unreliable, however, and disagrees in many respects with all others I have seen. He assigns dates to none of the events and discussions he describes, for many of which his is the only extant account, and his name does not appear anywhere else in the entire record of the Elizabeth construction works at Dover Harbor, so far as I am aware. I have therefore used his history sparingly.

78. "Articles to be considered of by the Commissioners for Dover Haven," 11 June 1584, SPD, 12/171/20.

79. The backwater pent became the Wellington Dock, the new harbor itself (originally dubbed Great Paradise) was later subdivided into Granville Dock and the Tidal Basin, and the long wall that separates them both from the sea is now the beachfront Esplanade, a main attraction in what has since become a seaside resort town as well as a busy port. Minet, "Some Unpublished Plans," 223–224. However, in the mid-1990s most of the area once occupied by the pent was filled in and transformed into a pedestrian retail space along the Esplanade.

80. Scot, "Dover haven," 867.

81. John Thomson to Thomas Cromwell, 17 October 1536, in *Letters and Papers*, vol. 11, no. 745 (17 October 1536). See also Biddle and Summerson, "Dover Harbour," 729–755.

82. Trew, in fact, was the only officer in the history of the project to receive an official commission of his own directly from the Privy Council and the lord chancellor. Even Digges held his office through the commission's appointment, though at the Privy Council's insistence.

83. Scott to Smith, 25 June 1584, SPD, 12/171/52.

THREE: Early Mathematical Navigation in England

1. Hugh Willoughby, "The true copie of a note found written in one of the two ships, to wit, the *Speranza*, which wintred in Lappia, where Sir Hugh Willoughby, and all his companie died, being frozen to death. Anno 1553," in *The Principall Navigations, Voiages and Discoveries of the English Nation* . . . , ed. Richard Hakluyt (London, 1589), 270, hereafter cited as *PN*.

2. John Janes, "The first voyage of Master Iohn Dauis . . . ," in *PN*, 776.

3. "A letter of the said M. Iohn Dauis, written to M. Sanderson of London, concerning his forewritten voyage," in *PN*, 792.

4. Steven Shapin and Simon Schaffer have argued that "technology" includes "literary and social practices, as well as . . . machines" and that "all three are *knowledge-producing tools*." *Leviathan and the Air-Pump: Hobbes, Boyle, and the Experimental Life* (Princeton, N.J.: Princeton University Press, 1985), 25 n (original emphasis). Likewise, my own use of the term *technologies* here refers to both the material "hardware" of navigational instruments and the mathematical "software" a pilot needed to understand and use them correctly.

5. J. A. Bennett has argued that mathematics in early modern England must be understood within a context of practical problem-solving; the "mathematical sciences" were explicitly perceived by their practitioners as a way to develop new solutions to pressing prob-

lems in navigation, surveying, cartography, and other valuable fields of practical endeavor. See Bennett, "Mechanics' Philosophy," 1–28; and Bennett, "Challenge of Practical Mathematics."

6. Spanish pilots of the sixteenth century were also resistant to many of the mathematical innovations urged upon them by the cosmographers of the Casa de Contratación; see Alison Sandman, "Mirroring the World: Sea Charts, Navigation, and Territorial Claims in Sixteenth-Century Spain," in *Merchants and Marvels: Commerce, Science, and Art in Early Modern Europe*, ed. Pamela H. Smith and Paula Findlen (New York: Routledge, 2002).

7. This was not true throughout Europe, however; Italian, Spanish, and Portuguese pilots developed and learned to rely upon mathematically based methods of navigation decades before their northern European counterparts. For a more detailed discussion of the late medieval and early modern European development of mathematical navigation technologies, see Eric H. Ash, "Navigation Techniques and Practices," in *The History of Cartography*, vol. 3, *Cartography in the European Renaissance and Reformation*, ed. David Woodward (Chicago: University of Chicago Press, expected 2005).

8. Mark Twain's *Life on the Mississippi* (1883; reprint, New York: Oxford University Press, 1996) provides a clear and engaging modern account of how the art of piloting is taught and learned through repetition and memorization. Samuel Clemens was trained as Mississippi riverboat pilot before his career as a writer.

9. For an example of a printed rutter including depictions of coastal landmarks, see Cornelis Antoniszoon, *The Safeguard of Sailers: or great Rutter . . .* , trans. Robert Norman, 2d ed. (London, 1587).

10. Waters, *Art of Navigation*, 78; see also David B. Quinn and A. N. Ryan, *England's Sea Empire, 1550–1642* (London: George Allen and Unwin, 1983), 19.

11. Quinn and Ryan, *England's Sea Empire*, 21. The literature on the medieval and early modern English wool and cloth trade is voluminous; for an overview, see Peter J. Bowden, *The Wool Trade in Tudor and Stuart England* (London: Macmillan, 1962); Peter Ramsey, *Tudor Economic Problems* (London: V. Gollancz, 1965); and Robert Brenner, *Merchants and Revolution: Commercial Change, Political Conflict, and London's Overseas Traders, 1550–1653* (Princeton, N.J.: Princeton University Press, 1993), pt. 1.

12. Waters, *Art of Navigation*, 83; Quinn and Ryan, *England's Sea Empire*, 21. Although a small number of English merchants had been interested in promoting overseas exploration for decades, they had always constituted a tiny minority, and they failed to attract much support for their schemes until the broader economic decline had set in. See E. G. R. Taylor, introduction to Roger Barlow, trans., *A brief summe of geographie*, Hakluyt Society Publications, 2d ser., no. 69 (London, 1932).

13. See "Robert Thorne's exhortation to Henry VIII to attempt discoveries northward," 1527, Lansdowne MSS, 100/7.

14. Martín Cortés, *The Arte of Nauigation . . .* , trans. Richard Eden (London, 1561), fol. A.iii v.

15. Tony Campbell, "Portolan Charts from the Late Thirteenth Century to 1500," in *The History of Cartography*, vol. 1, *Cartography in Prehistoric, Ancient, and Medieval Europe and the Mediterranean*, ed. J. B. Harley and David Woodward (Chicago: University of Chicago Press, 1987), 386; Waters, *Art of Navigation*, 67. Longitude, unmeasurable at sea until the mid–eighteenth century, was not depicted on early-sixteenth-century charts.

16. Waters, *Art of Navigation*, 83–84; Scott, *Constitution and Finance*, 1:18; Alison Sand-

man and Eric H. Ash, "Trading Expertise: Sebastian Cabot between Spain and England," *Renaissance Quarterly* 57 (2004): 813–46.

17. Roger Bodenham, "The voyage of M. Roger Bodenham with the great Barke Aucher to Candia and Chio, in the yeere 1550," in *The Principal Navigations, Voiages, Traffiques and Discoueries of the English Nation . . .* , ed. Richard Hakluyt, 2d rev. ed., 3 vols. (London, 1598–1600), 2:99–101, hereafter cited as *PN*2.

18. "The Charter of the Marchants of Russia, graunted vpon the discouerie of the saide Countrey, by King Philip and Queene Mary," in *PN*, 305. The company's official title upon its formal reestablishment by act of Parliament in 1566 was "The Fellowship of English Merchants for Discovery of New Trades." "Muscovy Company" and sometimes "Russia Company" were ubiquitous but informal references, and indeed the company was periodically involved in far more than just the Russia trade; see Scott, *Constitution and Finance*, 2:36–52.

19. "The Charter," in *PN*, 308.

20. Ibid., 305.

21. "The Testimonie of M. Richard Eden in his Decades, concerning this Booke," in *PN*, 270.

22. It should be remembered that the English merchants for whom Cabot wrote were as inexperienced at trading in distant overseas markets as English mariners were in sailing to them. For a discussion of Cabot's trade-related instructions, see Eric H. Ash, "'A Note and a Caveat for the Merchant': Mercantile Advisors in Elizabethan England," *Sixteenth Century Journal* 33 (2002): 1–31.

23. Sebastian Cabot, "Ordinances, instructions, and aduertisements of and for the direction of the intended voyage for Cathaye . . . ," in *PN*, 259.

24. Ibid.

25. Ibid., 263.

26. Ibid., 259.

27. Alison Sandman, letter to author, 5 September 2003.

28. For example, the instructions given to Anthony Jenkinson for his 1557 voyage to Russia stated that all ships were to follow the lead of the Admiral ship "and that no course nor weighing (in harbor especially) shall be made without the advice, consent and agreement of the said captain [Jenkinson], the master, his mate, and two other officers of the said ship, or of three of them at the least." Although Jenkinson's navigation was still a collaborative activity, the number of collaborators was less than half that of Willoughby's 1553 council, even with four ships to Willoughby's three; and all the officers included were limited to the Admiral ship, rather than being spread throughout the fleet. This was also more in keeping with contemporary Spanish practice. "Instructions giuen to the Masters and Mariners to be obserued in and about this Fleete, passing this yeere 1557 toward the Bay of S. Nicholas in Russia . . . ," in *PN*, 332.

29. Cabot, "Ordinances," in *PN*, 259.

30. Alison Sandman has studied the Spanish cartographers of the Casa de Contratación and their reliance upon navigational data provided by Iberian pilots in assembling their navigational charts. As is implied in Cabot's instructions, the Casa's cartographers believed that the most accurate charts could be compiled only through a broad, collaborative endeavor to obtain the most precise and comprehensive navigational data, a task that only the pilots of Spain's vast merchant fleets would ever be in a position to undertake. The cartographers

were repeatedly frustrated, however, by what they perceived as the pilots' lack of mathematical ability and careless imprecision in making their measurements. See Alison Deborah Sandman, "Cosmographers vs. Pilots: Navigation, Cosmography, and the State in Early Modern Spain," Ph.D. diss., University of Wisconsin–Madison, 2001, chap. 2.

31. Cabot, "Ordinances," in *PN*, 260.

32. Magnetic variation fluctuates (unpredictably) with respect to one's longitude and thus changes most rapidly at higher latitudes, as the longitude lines converge toward the poles.

33. Cabot, "Ordinances," in *PN*, 259. See Waters, *Art of Navigation*, appendixes 2 and 12, for transcriptions of sixteenth-century sources indicating that pilots and masters owned the navigational equipment they used; both sources describe the theft of pilots' personal equipment by pirates. It was also not uncommon, especially among English pirates and privateers, to kidnap the pilots themselves as well as stealing their charts and instruments, in order to take full advantage of their navigational expertise.

34. Robert Recorde, preface to *The Pathway to Knowledge* (London, 1551), fols. ζ.iv v–τ.i r.

35. Robert Recorde, *The Whetstone of Witte* (London, 1557), fol. A.iii r.

36. John Dee, *A Letter containing a most briefe Discourse Apologeticall* (London: 1592); printed in *Autobiographical Tracts of Dr. John Dee . . .* , ed. James Crossley, *Remains, Historical and Literary, Connected with the Palatine Counties of Lancaster and Chester* (Manchester: Chetham Society, 1851), 24:74–75. John Dudley, Duke of Northumberland and lord president of Edward VI's Privy Council, was the most prominent of Dee's patrons at this time; he was also a leading figure in bringing Sebastian Cabot back into English service.

37. John Dee, *General and Rare Memorials pertayning to the Perfect Arte of Navigation . . .* (London, 1577), fol. ε.iii r.

38. "The newe Nauigation and discouerie of the kingdome of Moscouia, by the Northeast, in the yeere 1553 . . . ," in *PN*, 280–281. The company's decision to appoint someone who was not an experienced mariner as the commanding officer of the fleet, especially when his height and social status were the major factors in their decision, may seem questionable. However, the captain general's chief role was to negotiate profitable trading relations with any peoples encountered in the journey, and it was therefore of paramount importance to have a man of ambassadorial rank and bearing in that office.

39. "A discourse of the honourable receiuing into England of the first Ambassador from the Emperor of Russia, in the yeere of Christ, 1556 . . . ," in *PN*, 321–322. The Russian ambassador, ironically, was one of the very few to survive the wreck, and once he finally reached London he was well received by company and court alike.

40. Willoughby, "The true copie," in *PN*, 268–269; Kenneth R. Andrews, *Trade, Plunder, and Settlement: Maritime Enterprise and the Genesis of the British Empire, 1480–1630* (Cambridge: Cambridge University Press, 1984), 67.

41. "The Nauigation and discouerie toward the River of Ob, made by Master Stephen Burrowe . . . 1556," in *PN*, 311. Exactly when Borough returned from his mission is not known, though he likely arrived at London in the autumn of 1557.

42. For further discussion of this technique and various contemporary approaches to it, see chapter 4, this volume.

43. "River of Ob," in *PN*, 314–315.

44. Ibid., 317–318.

45. "The voiage of the foresaid M. Stephen Burrough, An. 1557. from Colmogro to Wardhouse...," in *PN*, 326–331.

46. Richard Hakluyt, *Divers voyages touching the discouerie of America, and the Ilands adiacent unto the same*... (London, 1582), fol. ¶3 r.

47. Richard Eden had previously translated into English Sebastian Munster's *A treatyse of the newe India*... (London, 1553) and Peter Martyr's *The decades of the newe worlde or west India*... (London, 1555).

48. Richard Eden, preface to Cortés, *Arte of Nauigation*, fol. [C.iv r]–CC.i r.

49. Cortés's instructions for *using* the instruments once they were constructed, however, were often quite confusing; see chapter 4.

50. Waters, *Art of Navigation*, 102.

51. Stephen Borough, "Petition to create the office of chief pilot of England," [1562], Lansdowne MSS, 116/3, fol. 6 r; the petition has been transcribed and printed in Waters, *Art of Navigation*, appendix 6A, 513–514.

52. Borough, "Petition," Lansdowne MSS, 116/3, fol. 6 r–v.

53. Ibid., fol. 6 v–7 r.

54. Draft of letters patent naming Stephen Borough chief pilot of England, 3 January 1563, Lansdowne MSS, 116/3, fol. 4 r–v; printed in Waters, *Art of Navigation*, appendix 6B, 515–516.

55. Waters, *Art of Navigation*, 106 n.

56. Ibid., 106–113.

57. Michael Lok, "An account of Martin Frobisher's attempted voyages to reach Cathay," 1577, Cotton MSS, Otho E/VIII, fol. 48 r. This and many other manuscripts related to Frobisher's voyages have been transcribed and printed in *The Three Voyages of Martin Frobisher*..., ed. Richard Collinson, Hakluyt Society Publications, 1st ser., no. 38 (London, 1867).

58. "Examination of Martin Frobisher by Dr. Lewes on suspicion of preparing to commit piracy," 11 June 1566, SPD, 12/40/7; Captain Edward Horsey to Burghley, 21 August 1571, SPD, 12/80/31.

59. Lok, "Account of Frobisher's voyages," 1577, Cotton MSS, Otho E/VIII, fol. 43 r–v.

60. Ibid., fol. 43 r.

61. Michael Lok, "An account of the costs and expenses for Martin Frobisher's first two voyages," 1577, Lansdowne MSS, 24/62, fol. 163 r.

62. Lok, "Account of Frobisher's voyages," 1577, Cotton MSS, Otho E/VIII, fol. 42 r–v.

63. Lok, "Account of costs and expenses," 1577, Lansdowne MSS, 24/62, fol. 163 v.

64. Lok, "Account of Frobisher's voyages," 1577, Cotton MSS, Otho E/VIII, fol. 44 r; the manuscript has been damaged by fire, and there are some lacunae in the text as a result.

65. William Borough, "A dedicatorie Epistle vnto the Queenes most excellent Maiestie...," in *PN2*, 1:417–418.

66. Lok, "Account of Frobisher's voyages," 1577, Cotton MSS, Otho E/VIII, fol. 44 v.

67. Frobisher's fellow ship's master on the voyage, Christopher Hall, might have been another story. Although his experience at sea could not match Frobisher's, Hall sailed from the docks at Limehouse, just downriver from London, making it very likely that Stephen Borough had already overseen and approved at least some part of his navigational training. Ibid., fol. 43 r.

68. Michael Lok, "An account of expenses for the Company of Cathay, in preparing for

Frobisher's 1576 voyage," 1576, Additional MSS, 39852, fol. 20. This list of navigational equipment taken on a voyage of exploration in 1576 is very rare, and it is valuable for the insights it provides regarding the perceived "cutting edge" of the field. It must be stressed, however, that the list does not represent a standard complement of navigational equipment, as Frobisher's voyage was a far more ambitious and uncertain endeavor than a typical mercantile venture. For a discussion of Dee's development of the circumpolar projection, see Waters, *Art of Navigation*, 209–212. The circumpolar chart was often referred to as a "paradoxal chart" or "paradoxal compass" in sixteenth-century England because rhumb lines, depicted as straight lines on a conventional plane chart, were correctly manifested as the spiral lines they would be on a spherical surface. Dee claimed to have invented the circumpolar projection, but the Spanish had been experimenting with it for charting regions near the South Pole since Magellan's voyage of 1519–22, though Dee may have been unaware of the Spanish innovation.

69. Dee, *General and Rare Memorials*, 2–3.

70. George Best, *A true discourse of the late voyages of discouerie* . . . (London, 1578), 5 (original emphasis). Unfortunately, none of Frobisher's recorded observations survive.

71. William Borough was a strong advocate of recording and charting the magnetic variation at every opportunity, and he later wrote a treatise explaining in detail how to do it. See William Borough, *A Discovrs of the Variation of the Cumpas*, published with Robert Norman, *The Newe Attractiue* . . . (London, 1581), and the discussion of the text in chapter 4.

72. "A narrative of the unwarrantable doings of Captain Frobisher," [1578?], Lansdowne MSS, 100/1, fol. 8 v–9 r. Christopher Hall, as the master of another ship in Frobisher's fleet, pointed out Frobisher's mistake and refused to follow him into the bay. His willingness to risk a charge of mutiny in this case is a strong testament to his confidence in his own piloting abilities.

73. Dee, *General and Rare Memorials*, 3.

74. Willoughby, "The true copie," in *PN*, 266; Taylor, *Tudor Geography*, 126. Taylor's biographical notes must be treated with some caution here; she uncharacteristically confuses Arthur Pett the mariner with Peter Pett the shipwright, who was active in rebuilding Dover Harbor from 1579 onward (see chapter 2, this volume).

75. "Commission giuen . . . vnto Arthur Pet, and Charles Iackman, for a voyage by them to be made, for discouerie of Cathay, 1580 . . . ," in *PN*, 455.

76. Ibid., 455–457.

77. William Borough, "Instructions and notes very necessarie and needefull to be obserued, in the purposed voyage for discouerie of Cathay Eastwards, by Arthur Pet, and Charles Iackman," in *PN*, 458–459 (original emphasis). So concerned was Borough with data collection, in fact, that he used some variant of the word *note* in every single paragraph of his brief instructions, six times in the first paragraph alone, and a total of twenty-three times throughout.

78. Best, *A true discourse*, 5.

79. William Bourne, *A Regiment for the Sea* . . . , 3d ed. (London, 1580), fol. B.ii v.

FOUR: Secants, Sailors, and Elizabethan Manuals of Navigation

1. John Dee, "Mathematicall Praeface," in *The Elements of Geometrie of* . . . *Euclide*, trans. Henry Billingsley (London, 1570), fol. d.iiii v (original emphasis); see also the "Groundplat of my Mathematicall Praeface," bound after fol. A.iiii.

2. Dee, "Mathematicall Praeface," fol. d.iiii v.

3. On the training of early modern English seamen, see Kenneth R. Andrews, "The Elizabethan Seaman," *Mariner's Mirror* 68 (1982): 245–262.

4. The literature on the culture of early modern practical mathematics and its patronage is vast. For the Continental context, see Mario Biagioli, "The Social Status of Italian Mathematicians, 1450–1600," *History of Science* 27 (1989): 41–95; M. Henninger-Voss, "Working Machines and Noble Mechanics: Guidobaldo del Monte and the Translation of Knowledge," *Isis* 91 (2000): 233–259; Thomas B. Settle, "The Tartaglia-Ricci Problem: Towards a Study of the Technical Professional in the Sixteenth Century," in *Cultura, scienze e tecniche nella Venezia del cinquecento: Atti del convegno internazionale di studio Giovan Battista Benedetti e il suo tempo* (Venice: Istituto Veneto di Scienze, Lettere ed Arti, 1987); Grafton, *Leon Battista Alberti*, esp. chap. 3; Sandman, "Mirroring the World"; Robert S. Westman, "The Astronomer's Role in the Sixteenth Century: A Preliminary Study," *History of Science* 17 (1980): 105–147; and Bruce T. Moran, "German Prince-Practitioners: Aspects in the Development of Courtly Science, Technology, and Procedures in the Renaissance," *Technology and Culture* 22 (1981): 253–274.

5. Historiographical interest in the English context for practical mathematics and its patronage has also been especially intense; see Taylor, *Mathematical Practitioners*; Taylor, *Tudor Geography*; E. G. R. Taylor, *Late Tudor and Early Stuart Geography, 1583–1650* (London: Methuen, 1934); A. J. Turner, "Mathematical Instruments and the Education of Gentlemen," *Annals of Science* 30 (1973): 51–88; Bennett, "Mechanics' Philosophy," 1–28; Bennett, "Challenge of Practical Mathematics"; Johnston, "Mathematical Practitioners," 319–344; Johnston, "Making Mathematical Practice"; Frances Willmoth, *Sir Jonas Moore: Practical Mathematics and Restoration Science* (Woodbridge, Suffolk: Boydell Press, 1993); Frances Willmoth, "Mathematical Sciences and Military Technology: The Ordinance Office in the Reign of Charles II," in *Renaissance and Revolution: Humanists, Scholars, Craftsmen, and Natural Philosophers in Early Modern Europe*, ed. J. V. Field and Frank A. J. L. James (Cambridge: Cambridge University Press, 1993); Lesley B. Cormack, "Twisting the Lion's Tail: Practice and Theory at the Court of Henry Prince of Wales," in *Patronage and Institutions: Science, Technology, and Medicine at the European Court, 1500–1700*, ed. Bruce T. Moran (Rochester, N.Y.: Boydell Press, 1991); Lesley B. Cormack, *Charting an Empire: Geography at the English Universities, 1580–1620* (Chicago: University of Chicago Press, 1997), esp. chap. 3; Katherine Neal, "Mathematics and Empire, Navigation and Exploration: Henry Briggs and the Northwest Passage Voyages of 1631," *Isis* 93 (2002): 435–453; and Keith Thomas, "Numeracy in Early Modern England," *Transactions of the Royal Historical Society* 37 (1987): 103–132.

6. P. D. A. Harvey, *Maps in Tudor England* (London: Public Record Office and British Library, 1993); Peter Barber, "England I: Pageantry, Defense, and Government: Maps at Court," and "England II: Monarchs, Ministers, and Maps, 1550–1625," in *Monarchs, Ministers, and Maps: The Emergence of Cartography as a Tool of Government in Early Modern Europe*, ed. David Buisseret (Chicago: University of Chicago Press, 1992).

7. Feingold, *Mathematicians' Apprenticeship*, esp. chap. 6.

8. On the idea of an obligatory point of passage, see Bruno Latour, *The Pasteurization of France*, trans. by Alan Sheridan and John Law (Cambridge, Mass.: Harvard University Press, 1988); Bruno Latour, *Science in Action: How to Follow Scientists and Engineers through Society* (Cambridge, Mass.: Harvard University Press, 1987); and Bruno Latour, "Give Me a Lab-

oratory and I will Raise the World," in *Science Observed: Perspectives on the Social Study of Science*, ed. Karin Knorr-Cetina and Michael Mulkay (London: Sage, 1983).

9. The foundational works in this vein are Taylor's *Mathematical Practitioners* and *Tudor Geography*.

10. See, for example, J. A. Bennett, "Geometry and Surveying in Early 17th-Century England," *Annals of Science* 48 (1991): 345–354, on the mathematicians' struggle to introduce more complex mathematical instruments into the art of surveying and the practicing surveyors' resistance to them.

11. The best works illustrating the complexities and rivalries within the overall rubric of "mathematical practitioners" are Bennett, "Geometry and Surveying"; and Johnston, "Mathematical Practitioners." Both works demonstrate that social and intellectual advances for mathematicians often came at the expense of the more common sort of practitioner.

12. E. G. R. Taylor, editor's note to *A Regiment for the Sea, and other writings on navigation, by William Bourne of Gravesend, a Gunner (c. 1535–1582)*, Hakluyt Society Publications, 2d ser., no. 121 (Cambridge: Cambridge University Press, for the Hakluyt Society, 1963), 115 n.

13. William Bourne, *A Regiment for the Sea* . . . (London, [1574]), fol. A.iii v.

14. Ibid., [fol. A.iiii r].

15. Waters, *Art of Navigation*, 134, 143.

16. This technique was developed by the Portuguese during the fifteenth century, as they explored southward along the African coast and the Pole Star dropped below the horizon, creating the need for a new way to determine one's latitude from the stars. During the course of the solar year, the sun appears to travel along a circular path known as the ecliptic, which is set at an angle of roughly $23.5°$ from the celestial equator (and thus from the terrestrial one as well). The ecliptic intersects the equator at the two equinoxes and is most distant from it at the two solstices. As the year progresses, therefore, the sun wanders as far as $23.5°$ north of the equator (the summer solstice, in the Northern Hemisphere), crosses the equator moving southward (autumnal equinox), declines $23.5°$ to the south (winter solstice), and then crosses the equator again on its way north (vernal equinox). Because the calculation of latitude uses the sun as the only celestial reference point to determine the observer's angular distance from the equator, solar declination has an obvious impact on the calculation and must be added or subtracted appropriately. If the observer's latitude is higher than $23.5°$ (north or south), he simply combines the sun's declination with its altitude at noon (by addition or subtraction, depending on whether the sun is declined toward or away from the hemisphere in which the observer's ship happens to be). If the observer is within $23.5°$ of the equator, however, the calculation depends on whether his ship is closer to the equator than the sun is on the date in question.

17. Cortés, *Arte of Nauigation*, fol. 71 r–72 r.

18. Ibid.

19. The first example showed how to calculate the total yearly declination of the sun ($47°$), using the solar meridian altitudes taken on the summer and winter solstices at a given latitude. The second example did show how to determine the latitude of the observer, but only if the sun's meridian altitudes for the two solstices were known for that latitude. The third example explained how to determine the exact time when a solstice had occurred through interpolation, or the "rule of three." Each example required multiple observations to be made

from the same latitude on different dates—hardly feasible for anyone traveling at sea—and none directly addressed the immediate problem of finding one's present latitude.

20. Bourne, *Regiment* [1574], chaps. 7–10, fols. 29 r–36 r.

21. Chapter 7 covered sailing in the Northern Hemisphere, when the sun was to the south at noon; chapter 8 discussed sailing near the equator, when because of the sun's declination the observer's ship might be at a latitude between the equator and the sun; chapter 9 covered sailing in the Southern Hemisphere, with the sun to the north at noon; and the tenth chapter gave special rules for sailing above the Arctic Circle—a particularly English concern—which allowed two measurements of the sun to be used in the summer (when the sun visibly crossed the observer's meridian twice, at noon and at midnight).

22. Waters, *Art of Navigation*, 132.

23. Bourne, *Regiment* [1574], fol. 30 v–32 r.

24. Ibid.

25. Bourne, *Regiment* (1580), fol. B v–B.ii r.

26. Bourne, *Regiment* [1574], fol. 12 v. See, for example, Bourne's handling of lunar auge, or the moon's tendency to change velocity during the course of its orbit, in his method for calculating the time of high tide. After taking the trouble to introduce the phenomenon and explain why it occurs, Bourne then recommended to his readers that they simply ignore it, as it would have a negligible impact upon their calculations.

27. Ibid., fol. 58 r.

28. Borough, *Discovrs*, fol. *ii r–*iii r.

29. Ibid., fol. *ii v–*iii v.

30. Ibid., fol. Dii r–Diii r (original emphasis).

31. Richard Hakluyt to Francis Walsingham, 1 April 1584, SPD, 12/170/1.

32. The Thomas Smith in question was almost certainly the son of the royal customer of London of the same name, who had leased the mining rights of the Company of Mines Royal and acted as the treasurer of the Dover Harbor project. The younger Smith was an extraordinarily active London merchant; he served several terms as a governor of the Muscovy, East India, and North-West Passage companies, as well as the treasurer for the company administering the first Virginia settlement; see *Dictionary of National Biography*, s.v. "Thomas Smith"; and Waters, *Art of Navigation*, 320.

33. Hood had apparently sought to make the lectureship a permanent, endowed institution, but he was unsuccessful in securing the long-term support of his patrons among the Privy Council and the governors of the city of London, despite his reappointment in 1590. Indeed, he had some difficulty in collecting his salary during his tenure as lecturer. Hood to Burghley, [1588?], Lansdowne MSS, 101/12. For an analysis of the uneven progress of Hood's mathematical career in London, see Johnston, "Mathematical Practitioners," 330–340. My own interpretation of the increasing prominence of mathematical navigation owes much to Johnston's discussion of practical mathematicians such as Hood and their use of instruments as a means of raising their social status.

34. See Thomas Hood, *A Copie of the Speache: Made by the Mathematicall Lecturer, unto the Worshipfull Companye present* (London, 1588), fol. A.ii v–A.iii v. Unlike the other large nations of western Europe in the sixteenth century, England did not maintain a standing army, relying instead upon a formalized citizen militia for defense, the proper training of which was thus a critical concern as open hostilities with Spain commenced in 1588.

35. Waters, *Art of Navigation*, 185–186; Johnston, "Mathematical Practitioners," 334.

36. Thomas Hood, *The Vse of the Celestial Globe in Plano*... (London, 1590); Thomas Hood, *The Vse of Both the Globes, Celestiall, and Terrestriall*... (London, 1592); Thomas Hood, *The vse of the two Mathematicall Instrumentes*... (London, 1596); Thomas Hood, *The Marriners guide*..., appended to William Bourne, *A Regiment for the Sea*..., ed. Thomas Hood, 5th ed. (London, 1592); Christian Urstitius, *The Elements of Arithmeticke most methodically deliuered*..., trans. Thomas Hood (London, 1596); Peter Ramus, *Elementes of Geometrie*..., trans. Thomas Hood (London, 1590).

37. Hood, *Marriners guide*, fol. 4 v.

38. Ibid., fol. Aiii r (emphasis added).

39. Thomas Hood, dedication to Sir John Harte of Ramus, *Elementes*.

40. Hood, *Marriners guide*, fol. 4 r–5 r.

41. Johnston, "Mathematical Practitioners," 336–337; Jardine and Grafton, "'Studied for Action,'" 30–78; Sherman, *John Dee*.

42. Thomas Hood, *Vse of Both the Globes*.

43. Bourne, *Regiment* [1574], [fol. Q.iiii r–v].

44. John Davis, *The Seamans Secrets*... (London, 1595).

45. Ibid., fol. ¶3 v–[¶4 r].

46. Ibid., fol. [A4 v], B v, B3 v, and C3 v. The horizontal tide table, referred to by Davis on fol. B v, is missing in extant copies.

47. Ibid., unlabeled folio after [C4].

48. Ibid., fols. A3 v, B r, C3 v, K3 r.

49. Waters, *Art of Navigation*, 220.

50. Edward Wright, *Certaine Errors in Navigation* (London, 1599). Wright described the history of the work in his preface.

51. Ibid., fol. ¶¶¶ r–v.

52. Ibid., fol. [¶¶¶4 r].

53. Ibid., chaps. 5–6.

54. Ibid., fol. ¶¶¶2 v.

55. Ibid., fol. O r–v.

56. It should be noted that the true bearing between two points on a spherical surface and the constant bearing or rhumb line heading between them are not the same thing. The rhumb line depicts a single, constant compass heading and is therefore the *simplest* route to sail, but it actually traces a spiral rather than a straight line. This is because the angle of intersection with longitude lines is always maintained, even though the longitude lines themselves are converging. A great circle route, in contrast, is a straight line along the Earth's surface and is thus almost always shorter in linear distance than a rhumb line route, but it involves constant adjustment of one's compass heading so as to avoid veering toward the pole. The Mercator projection chart made rhumb-line navigation (using a single compass heading) simpler and more accurate than was possible on a plane chart but did not provide the shortest or most direct route possible because it did nothing to simplify great circle navigation.

57. In practice, when mariners actually tried to navigate in higher latitudes using the maps, they were often disappointed with the results. The inaccuracy of many geographical positions placed on even a correctly proportioned Mercator projection chart led to serious

navigational errors and initially retarded the popularity of the new charts. See Waters, *Art of Navigation*, 223–224.

58. Thomas Harriot, *A briefe and true report of the new found land of Virginia* . . . (London, 1588). Unlike most of the authors considered in this chapter, a considerable historiography has been devoted to Thomas Harriot. For an excellent and comprehensive biography, see John W. Shirley, *Thomas Harriot: A Biography* (Oxford: Clarendon Press, 1983). For a series of essays that address separately the diverse facets of Harriot's intellectual endeavors, see *Thomas Harriot: Renaissance Scientist* (Oxford: Clarendon Press, 1974)—on navigation specifically, see Jon V. Pepper's essay, "Harriot's Earlier Work on Mathematical Navigation: Theory and Practice."

59. Thomas Harriot, "Instructions for Raleigh's voyage to Guyana," [1595], Additional MSS, 6788, fol. 486 r–487 r. Harriot's mathematical papers were not arranged and bound in order, and the six lectures are now somewhat scattered between Additional MSS 6788 and 6789; for a detailed account of their location and complete contents, see Pepper, "Harriot's Earlier Work," 57–75.

60. Harriot, "Instructions," fol. 487 r–488 r.

61. Ibid., fol. 489 r.

62. Wright, *Certaine Errors*, fol. N3 v–O r.

63. Ibid., fol. O2 r–v.

64. Harriot, "Instructions," 473 r (emphasis added).

65. Ibid., fol. 474 r; the diagram is transcribed in Pepper, "Harriot's Earlier Work," 66.

66. Harriot, "Instructions," fol. 475 r; Pepper has transcribed Harriot's second example for this lecture. "Harriot's Earlier Work," 67.

67. Thomas Blundeville, *M. Blvndevile, His Exercises* . . . (London, 1594), unlabeled folio after A4. Elizabeth Bacon was a very well-connected young woman: she was the wife of Francis Wyndham, a judge of the Court of Common Pleas; the daughter of Nicholas Bacon, Queen Elizabeth's lord keeper of the great seal; and the sister of Francis Bacon, later lord chancellor under James I, a lifelong advocate for the reform of natural philosophy and the subject of the following chapter.

68. Ibid., unlabeled folio after A4.

69. Ibid., title page.

70. Ibid., fol. A3.

71. Not all of Blundeville's sources had been published by 1594, however. For example, he included a brief description of Wright's Mercator projection chart and his method for constructing it (with Wright's permission) five years before Wright published it himself in his *Certaine Errors*.

72. Blundeville, *Exercises*, chap. 47.

73. Ibid., fol. A4 v.

74. Johnston, "Mathematical Practitioners," 320.

75. Edward Worsop, *A Discoverie of sundrie errours and faults daily committed by Landemeaters* . . . (London, 1582), fol. F3 r, F2 v, and E4 r.

76. Ibid., fol. F2 v and G1 r.

77. Roger Ascham, *The Scholemaster* . . . (London, 1570), fol. 5 v (emphasis added).

78. Gilbert, "The erection of an Achademy," fols. 3 r–4 r and 6 v.

79. Robert Norman, *The Newe Attractiue* . . . (London, 1581), fol. B.i r–v.

80. William Borough, "A dedicatorie Epistle vnto the Queenes most excellent Maiestie . . . ," in *The Principal Navigations, Voiages, Traffiques and Discoueries of the English Nation . . .* , ed. Richard Hakluyt, 2d rev. ed., 3 vols. (London, 1598–1600), 1:417–418.

81. Digges and Digges, *Stratioticos*, [fol. A.iiii r–v]. Interestingly, although Digges was emphatic that his brief experience at sea confirmed the existence of sundry errors in navigational practice, he did not claim that it demonstrated the efficacy of his own mathematically based solutions.

82. See Taylor, *Mathematical Practitioners*, 173–175; Johnston, "Making Mathematical Practice," chap. 2.

83. Neal, "Mathematics and Empire," provides an excellent example of how two navigators (Thomas James and Luke Foxe) could disagree strongly with respect to the importance of navigational theory versus practice at sea and yet still be part of "the same navigational culture" (453), allowing them to communicate with each other effectively enough to disagree in the first place.

84. Bennett, "Challenge of Practical Mathematics," and Johnston, "Mathematical Practitioners," appear to make similar arguments, but they draw conclusions that are in some ways the opposite of my own. Both authors describe a separate class of practical mathematical disciplines ("mathematical sciences" for Bennett, "the mathematicalls" for Johnston) that were prevalent in early modern England, defined in terms of similar problems, practices, and technologies. Moreover, both authors do much to illustrate the complex interactions and rivalries that could emerge within the broad community of mathematical practitioners they seek to define and defend. Each remains committed, however, to postulating a community of "self-styled mathematical practitioners," which they portray as being coherent and recognizable to both members and nonmembers. Johnston, "Mathematical Practitioners," 319. Without denying the mathematical links between the individuals they describe, I would argue that their integrity as a "self-styled" community with common interests has been overstated and that this has worked to obscure the multifaceted interrelationships that are the most interesting part of each author's work.

85. This shift in pilots' notion of what constituted "traditional" practice was similar in Spain; see Sandman, "Cosmographers vs. Pilots," 289–290.

FIVE: Francis Bacon and the Expertise of Natural Philosophy

1. Richard Yeo, "An Idol of the Marketplace: Baconianism in Nineteenth-Century Britain," *History of Science* 23 (1985): 251–298; Benjamin Farrington, *Francis Bacon: Philosopher of Industrial Science* (New York: H. Schuman, 1949); Benjamin Farrington, *Francis Bacon: Pioneer of Planned Science* (New York: Praeger, 1963); Robert K. Faulkner, *Francis Bacon and the Project of Progress* (Lanham, Md.: Rowman and Littlefield, 1993); Julie Robin Solomon, *Objectivity in the Making: Francis Bacon and the Politics of Inquiry* (Baltimore: Johns Hopkins University Press, 1998).

2. Antonio Pérez-Ramos, "Bacon's Legacy," in *The Cambridge Companion to Bacon*, ed. by Markku Peltonen (Cambridge: Cambridge University Press, 1996), 316. Pérez-Ramos's essay analyzes various impressions and manifestations of Bacon's philosophy and philosophical method from his death through the twentieth century. See also Steven Shapin, *A Social History of Truth: Civility and Science in Seventeenth-Century England* (Chicago: Univer-

sity of Chicago Press, 1994); Steven Shapin, *The Scientific Revolution* (Chicago: University of Chicago Press, 1996); William Eamon, *Science and the Secrets of Nature: Books of Secrets in Medieval and Early Modern Culture* (Princeton, N.J.: Princeton University Press, 1994); and Christopher Hill, *Intellectual Origins of the English Revolution* (Oxford: Clarendon Press, 1965), 125–130. Hill also traces the threads of Bacon's thought through the decades before and during the English Civil War, prior to the official founding of the Royal Society.

3. Thomas Sprat, *The History of the Royal-Society of London, For the Improving of Natural Knowledge* (London, 1667), 35–36 (original emphasis).

4. Pérez-Ramos, "Bacon's Legacy," 314–319.

5. Robert Hooke, preface to *Micrographia* (London, 1665).

6. Pérez-Ramos, "Bacon's Legacy," 311–312, 327–330.

7. Julian Martin, *Francis Bacon, the State, and the Reform of Natural Philosophy* (Cambridge: Cambridge University Press, 1992); John E. Leary Jr., *Francis Bacon and the Politics of Science* (Ames: Iowa State University Press, 1994); Stephen Gaukroger, *Francis Bacon and the Transformation of Early-Modern Philosophy* (Cambridge: Cambridge University Press, 2001); Deborah E. Harkness, "'Strange' Ideas and 'English' Knowledge: Natural Science Exchange in Elizabethan London," in *Merchants and Marvels: Commerce, Science, and Art in Early Modern Europe*, ed. Pamela H. Smith and Paula Findlen (New York: Routledge, 2002), 137–160. In somewhat earlier work, Paolo Rossi has done an excellent job of connecting Bacon with the sixteenth century's renewed interest in the mechanical arts and the role they could potentially play in the pursuit of natural philosophical knowledge, not just in England but throughout western Europe; see his books *Francis Bacon: From Magic to Science*, trans. Sacha Rabinovitch (Chicago: University of Chicago Press, 1968); and *Philosophy*.

8. My brief summary of Bacon's natural philosophy is based in part on Paolo Rossi, "Bacon's Idea of Science," in the *Cambridge Companion to Bacon*; Antonio Pérez-Ramos, *Francis Bacon's Idea of Science and the Maker's Knowledge Tradition* (Oxford: Clarendon Press, 1988); Brian Vickers, "Francis Bacon and the Progress of Knowledge," *Journal of the History of Ideas* 53 (1992): 495–518; Martin, *State*; and Gaukroger, *Transformation*.

9. Francis Bacon, *New Organon* (London, 1620), aph. 83; trans. and ed. James Spedding, Robert Leslie Ellis, and Douglas Denon Heath, in *The Works of Francis Bacon* (Stuttgart-Bad Cannstatt: Friedrich Frommann Verlag Gunther Holzboog, 1961–63), 4:81. The publication information for modern editions of Bacon's collected works can be confusing and requires some explanation. The Spedding-Ellis-Heath edition of *The Works of Francis Bacon* was first published in London in seven volumes, 1857–61, followed by *The Letters and Life of Francis Bacon* in seven more volumes, edited by James Spedding alone, 1861–74. These two London editions have been combined into a single fourteen-volume facsimile reprint, published in Stuttgart from 1961 through 1963; this is the edition cited above. This reprint should not be confused with various nineteenth-century American editions of the *Works*, which include fifteen volumes and have a different pagination. All references to Bacon's works in this chapter refer to the fourteen-volume 1961–63 Stuttgart reprint edition and are cited hereafter as "Spedding," followed by the relevant volume and page numbers; where applicable (in the *New Organon* and "Parasceve"), I have also provided the number of the aphorism in question. All translations are likewise taken from these volumes.

10. William R. Newman, however, has written about an alternative Aristotelian school of thought, prevalent in some alchemical circles, which did not insist upon a strict philo-

sophical distinction between natural and artificial knowledge; see the first chapter of *The Summa Perfectionis of Pseudo-Geber: A Critical Edition, Translation, and Study* (Leiden: E. J. Brill, 1991).

11. Francis Bacon, preface to *Great Instauration* (London, 1620), in Spedding, 4:14.

12. Francis Bacon, "Parasceve" (London, 1620), aph. 1, in Spedding, 4:253.

13. Ibid., aph. 5, in Spedding, 4:257.

14. Lorraine Daston and Katherine Park, *Wonders and the Order of Nature, 1150–1750* (New York: Zone Books, 1998), chaps. 6 and 7; Horst Bredekamp, *The Lure of Antiquity and the Cult of the Machine: The Kunstkammer and the Evolution of Nature, Art, and Technology*, trans. Allison Brown (Princeton, N.J.: Markus Wiener, 1995), 63–80; and Thomas DaCosta Kaufmann, *The Mastery of Nature: Aspects of Art, Science, and Humanism in the Renaissance* (Princeton, N.J.: Princeton University Press, 1993), 184–194.

15. Bacon, *New Organon*, aph. 81, in Spedding, 4:79 (emphasis added).

16. Ibid., aph. 73, in Spedding, 4:73.

17. Bacon, "Parasceve," aph. 5, in Spedding, 4:258.

18. Bacon, *New Organon*, aph. 117, in Spedding, 4:104.

19. Bacon, "Parasceve," preface, in Spedding, 4:251–252.

20. Bacon, *New Organon*, aph. 70, in Spedding, 4:70.

21. Ibid., aph. 61 in Spedding, 4:62–63.

22. William Rawley, "The Life of the Right Honourable Francis Bacon, Baron of Verulam, Viscount St. Alban," appended to *Resuscitatio* (London, 1657), in Spedding, 1:4 (emphasis added). When Bacon left Cambridge in 1575, he was still fifteen years old.

23. Farrington, *Philosopher*, 14; Rossi, *Philosophy*, 2–3. See also *The Admirable Discourses of Bernard Palissy*, trans. Aurèle La Rocque (Urbana: University of Illinois Press, 1957).

24. Leon Battista Alberti, *On Painting*, trans. Cecil Grayson (London: Penguin Books, 1991); Leon Battista Alberti, *On the Art of Building in Ten Books*, trans. Joseph Rykwert, Neil Leach, and Robert Tavernor (Cambridge, Mass: MIT Press, 1988); Filarete (Antonio Averlino), *Treatise on Architecture: Being the Treatise by Antonio di Piero Averlino, Known as Filarete*, ed. and trans. John R. Spencer, 2 vols. (New Haven, Conn.: Yale University Press, 1965); Agricola, *De re metallica*; Vannoccio Biringuccio, *The Pirotechnia of Vannoccio Biringuccio: The Classic Sixteenth-Century Treatise on Metals and Metallurgy*, trans. Cyril Stanley Smith and Martha Teach Gnudi (New York: Dover, 1990); Taccola (Mariano di Jacopo), *De machinis: The Engineering Treatise of 1449*, ed. and trans. Gustina Scaglia, 2 vols. (Wiesbaden: Reichert, 1971); Ramelli, *Various and Ingenious Machines*.

25. Farrington, *Philosopher*, 12.

26. Gaukroger, *Transformation*, 37–67 and 155–160.

27. Spedding, 8:107; Leary, *Politics of Science*, 25–37; Lisa Jardine and Alan Stewart, *Hostage to Fortune: The Troubled Life of Francis Bacon* (London: Victor Gollancz, 1998), 124, 140–144. Burghley's reluctance to help his nephew is curious, but it probably resulted from his desire to advance his own son, Robert Cecil, with whom Bacon would have been in competition. Martin, however, has argued that although Bacon did not succeed in fulfilling all of his grand ambitions during Elizabeth's reign, this was mostly due to his relative lack of experience, and that his patronage record during the sixteenth century must be seen as moderately successful for someone in his circumstances. *State*, 23–44.

28. Bacon to Burghley, [1592], in Spedding, 8:108–109 (original emphasis); see also Martin, *State*, 60–64.

29. Bacon's request to Burghley for patronage may be compared with John Dee's numerous contemporary (and equally unsuccessful) attempts to gain royal support for his philosophical endeavors, especially in subsidizing and expanding what he termed his "*Mortlacensi Hospitali Philosophorum peregrinantium.*" See John Dee, "The Compendious Rehearsal," printed in *Autobiographical Tracts of Dr. John Dee . . .* , ed. James Crossley, *Remains, Historical and Literary, Connected with the Palatine Counties of Lancaster and Chester* (Manchester: Chetham Society, 1851), 24:40.

30. Martin, *State*, 64–69; Gaukroger, *Transformation*, 70–73; Spedding, 8:119–123. The entertainment also included speeches praising fortitude, love, and finally Elizabeth herself as the embodiment of all the virtues.

31. Francis Bacon, "In Praise of Knowledge," c. 1592, in Spedding, 8:123–124.

32. Ibid., 8:124.

33. Spedding, 8:325–329, 342; Martin, *State*, 69–71; Gaukroger, *Transformation*, 72–73.

34. Francis Bacon, "The Second Councilor, advising the Study of Philosophy," 1594, in Spedding, 8:334–335. For further consideration of Bacon's speeches at the 1594 *Gesta Grayorum* and their relation to the *Kunstkammer* tradition, see Daston and Park, *Wonders*, 290–296; and Kaufmann, *Mastery of Nature*, 184–194.

35. Francis Bacon, *Of the Advancement of Learning* (London, 1605), in Spedding, 3:294.

36. Ibid., 3:330.

37. Ibid., 3:332.

38. Ibid., 3:332–333.

39. Bourne, *Regiment* [1574], fol. [A.iv r].

40. Blundeville, *Exercises*, fol. A3 v.

41. Sebastian Cabot, "Ordinances, instructions, and aduertisements of and for the direction of the intended voyage for Cathaye . . . ," in *The Principall Navigations, Voiages and Discoveries of the English Nation . . .* , ed. Richard Hakluyt (London, 1589; hereafter *PN*), 259.

42. William Borough, "Instructions and notes very necessarie and needefull to be obserued, in the purposed voyage for discouerie of Cathay Eastwards, by Arthur Pet, and Charles Iackman," in *PN*, 458–459; John Dee, "Certaine briefe aduises, giuen by Master Dee, to Arthur Pet, and Charles Iackman, to be obserued in their Northeasterne discouerie, Anno 1580," in *PN*, 459.

43. Bacon, *New Organon*, aph. 74, in Spedding, 4:74–75; Bacon, "Parasceve," preface, in Spedding, 4:251–252.

44. Borough, "Petition" [1562], Lansdowne MSS, 116/3.

45. Richard Hakluyt to Francis Walsingham, 1 April 1584, SPD 12/170/1; Hood, *Copie of the Speache*; Hood to Burghley, [1588?], Lansdowne MSS, 101/12.

46. Hood, *Marriners guide*, fol. Aiii r and 5 r.

47. Davis, *Seamans Secrets*, fol. ¶3 r.

48. Borough, "dedicatorie Epistle," in *The Principal Navigations, Voiages, Traffiques and Discoueries of the English Nation . . .* , ed. Richard Hakluyt, 2d rev. ed., 3 vols. (London, 1598–1600), 1:417.

49. Digges and Digges, *Stratioticos*, [fol. A.iiii r–v].

50. Bourne, *Regiment* (1580), fol. B v–B.ii r; emphasis added.

51. Borough, *Discovrs*, fol. *iii v.

52. Wright, *Certaine Errors*, fol. ¶¶¶ v.

53. Early modern belief in the joint importance of theoretical knowledge and practical

experience was not limited to the mathematical arts. Jean Bodin, in the dedication to Jean Tessier of his 1566 treatise *Method for the Easy Comprehension of History*, argued strongly that abstract knowledge of legal precepts, without any experience in applying them to forensic practice, was worse than useless: "Indeed, those who think they have acquired a knowledge of law without forensic training are obviously like men who have exercised constantly in a gymnasium, yet have never seen the line of battle and have never undergone the fatigue of military service." Jean Bodin, *Method for the Easy Comprehension of History*, trans. Beatrice Reynolds (New York: Columbia University Press, 1945), 4–5.

54. Bacon, *New Organon*, aph. 61, in Spedding, 4:62–63; Rossi, *Francis Bacon*, 33.

55. Martin, *State*, 105–140; Leary, *Politics of Science*, 145–150.

56. I am indebted to an anonymous reviewer for the Johns Hopkins University Press for this suggestion.

57. Francis Bacon, preface to "Of the Interpretation of Nature" (1603), in Spedding, 10:87.

58. Bacon, "Parasceve," preface, in Spedding, 4:251–252.

59. Ibid., 4:252.

60. Rose-Mary Sargent, "Bacon as an Advocate for Cooperative Scientific Research," in the *Cambridge Companion to Bacon*, 146–171; Leary, *Politics of Science*, 231–263; Martin, *State*, 134–140.

61. Francis Bacon, *New Atlantis* (London, 1627), in Spedding, 3:132–135. Subsequent citations are parenthetical in the text.

62. Bacon, *Advancement of Learning*, in Spedding, 3:294.

63. Faulkner has suggested, however, that Bacon never intended to expand *New Atlantis* beyond its published form. He argues that the extant end of the narrative, "while abrupt, certainly seems to fit" and that Bacon had a hand in editing the text and supervising its translation into Latin before his death, thereby granting it his implicit approval for publication as it then stood. *Project of Progress*, 233–236.

64. On the fruitful interactions among practitioners of several practical arts in early modern London, see Harkness, "'Strange' Ideas."

65. Martin, *State*, 134–140.

CONCLUSION: Power, Authority, and the Expert Mediator

1. I am indebted to an anonymous reviewer for the Johns Hopkins University Press for suggesting this point.

2. L. E. Harris, *Vermuyden and the Fens: A Study of Sir Cornelius Vermuyden and the Great Level* (London: Cleaver-Hume Press, 1953); H. C. Darby, *The Draining of the Fens*, 2d ed. (Cambridge: Cambridge University Press, 1956); Willmoth, *Sir Jonas Moore*; Keith Lindley, *Fenland Riots and the English Revolution* (London: Heinemann Educational Books, 1982).

# Essay on Sources

The archival research for this book was conducted almost entirely in two manuscript collections: the State Papers, Domestic Series, for the reign of Elizabeth I, located at the Public Record Office in London; and the first part of the Lansdowne Manuscripts (vols. 1–122) at the British Library, also in London. Both collections have been microfilmed (though the originals of the Lansdowne collection are also available for researchers) and are accessible at several research libraries in the United States and the United Kingdom. John Roche Dasent, ed., *Acts of the Privy Council of England*, n.s. (London: H. M. Stationery Office, 1890–1949), also provided supplemental information on the activities of the Elizabethan government relating to the projects in question. Other primary source research was conducted using the Early English Books collection, based in part on the catalog compiled by A. W. Pollard and G. R. Redgrave, *A Short-Title Catalogue of Books Printed in England, Scotland, and Ireland and of English Books Printed Abroad, 1475–1640*, 2d rev. ed. (London: Bibliographical Society, 1976–91). This invaluable collection, which includes images of the full text of every book included in the catalog, has long been available on microfilm in numerous research libraries and is currently being digitized for access on-line. Of particular importance for this book were Richard Hakluyt, ed., *The Principall Navigations, Voiages and Discoveries of the English Nation . . .* (London, 1589); and his second revised edition, titled *The Principal Navigations, Voiages, Traffiques and Discoueries of the English Nation . . .* and published in three volumes from 1598 through 1600. Through his timely editorial efforts, Hakluyt preserved hundreds of pages of documents relating to English exploration, trade, and navigational practice through the sixteenth century, most of which would otherwise have been lost to history; he has earned the enduring gratitude of this historian and many others. Finally, for Francis Bacon's works and correspondence, I have relied upon *The Works of Francis Bacon*, ed. and trans. James Spedding, Robert Leslie Ellis, and Douglas Denon Heath, 14 vols. (Stuttgart-Bad Cannstatt: Friedrich Frommann Verlag Gunther Holzboog, 1961–63).

Concerning the techniques, capabilities, and common practices of early modern copper miners, assayers, and smelters, William Barclay Parsons, *Engineers and Engineering in the Renaissance* (Baltimore: Williams and Wilkins, 1939), remains an essential source, supplemented by C. N. Bromehead, "Mining and Quarrying to the Seventeenth Century," and R. J. Forbes, "Metallurgy," both essays in *A History of Technology*, vol. 2, *The Mediterranean Civilizations and the Middle Ages, c. 700 BC to c. 1500 AD*, ed. Charles Singer, E. J. Holmyard, A. R. Hall, and Trevor I. Williams (Oxford: Clarendon Press, 1956). Georgius Agricola, *De re metallica* (Basil, 1556); trans. Herbert Clark Hoover and Lou Henry Hoover (1912; reprint, New York: Dover, 1950), is the best contemporary source; see also Bern Dibner, *Agricola on Metals* (Norwalk, Conn.: Burndy Library, 1958), for a helpful commentary. Pamela O. Long, "The Openness of Knowledge: An Ideal and Its Context in 16th-Century

Writings on Mining and Metallurgy," *Technology and Culture* 32 (1991): 318–355, not only contains much information regarding early modern mining texts and their patronage and authorship but also provides a most thorough bibliography for the subject of early modern mining in its footnotes. On the genesis, activities, and travails of the Company of Mines Royal and the Company of Mineral and Battery Works, see especially the works of M. B. Donald, *Elizabethan Copper: The History of the Company of Mines Royal, 1568–1605* (London: Pergamon Press, 1955); and *Elizabethan Monopolies: The History of the Company of Mineral and Battery Works from 1565 to 1604* (Edinburgh: Oliver and Boyd, 1961). W. G. Collingwood, ed. and trans., *Elizabethan Keswick: Extracts from the Original Account Books, 1564–1577, of the German Miners, in the Archives of Augsburg*, tract series, no. 8 (Kendal, U.K.: Titus Wilson, 1912), was most helpful in filling out the German investors' perspective. The three-volume work of William Robert Scott, *The Constitution and Finance of English, Scottish, and Irish Joint-Stock Companies to 1720* (1910–12; reprint, Gloucester, Mass.: P. Smith, 1968), remains a vital source of information on both of these (and many other) early English joint-stock companies. For an analysis of the lawsuit between the queen and the Earl of Northumberland concerning mining rights on private lands, a pivotal event in the company's early development, see Eric H. Ash, "Queen v. Northumberland, and the Control of Technical Expertise," *History of Science* 39 (2001): 215–240.

The rebuilding of Dover Harbor was an immensely complex undertaking, especially by sixteenth-century English standards—it has been unflatteringly characterized as "the Elizabethan equivalent of the Channel Tunnel project" (Patrick Collinson, "The Monarchical Republic of Elizabeth I," *Bulletin of the John Rylands University Library of Manchester* 69 [1986]: 418)—and the various narrative histories of the project reflect this. Beyond the manuscript record, I have relied upon one contemporary and three modern published accounts: Reginald Scot, "The note of Reginald Scot esquier concerning Dover haven," in *Holinshed's Chronicles of England, Scotland, and Ireland . . .* , ed. John Hooker (1586; reprint, London: J. Johnson et al., 1808), 4:845–868; William Minet, "Some Unpublished Plans of Dover Harbour," *Archaeologia* 72 (1922): 185–224; Martin Biddle and John Summerson, "Dover Harbour," in *The History of the King's Works*, ed. H. M. Colvin, John Summerson, Martin Biddle, J. R. Hale, and Marcus Merriman (London: H. M. Stationery Office, 1982), 4:729–768; and Stephen Johnston, "Making Mathematical Practice: Gentlemen, Practitioners, and Artisans in Elizabethan England" (D.Ph. diss., Cambridge University, 1994), chapter 5. In addition to these, Petra J. E. M. van Dam, "Ecological Challenges, Technological Innovations: The Modernization of Sluice Building in Holland, 1300–1600," *Technology and Culture* 43 (2002): 500–520, provides a good introduction to the basic principles of sixteenth-century Dutch hydraulic construction techniques, upon which the Dover project was partially based.

On early English navigation, in terms of both theory and practice, the indispensable starting point is still David W. Waters, *The Art of Navigation in England in Elizabethan and Early Stuart Times* (London: Hollis and Carter, 1958). E. G. R. Taylor, *The Haven-Finding Art: A History of Navigation from Odysseus to Captain Cook* (London: Hollis and Carter, 1958), is another useful, though more general, source. On the training of sixteenth-century mariners and officers, English and otherwise, see Kenneth R. Andrews, "The Elizabethan Seaman," *Mariner's Mirror* 68 (1982): 245–262; and Pablo E. Pérez-Mallaína, *Spain's Men of the Sea: Daily Life on the Indies Fleets in the Sixteenth Century*, trans. Carla Rahn Phillips (Baltimore: Johns Hopkins University Press, 1998). The work of Alison Sandman has been most helpful for this book, in illuminating the connections, similarities, and differences between English and Spanish navigational knowledge, training, and practice during the sixteenth century; see her article "Mirroring the World: Sea Charts, Navigation, and Territorial Claims in Sixteenth-Century Spain," in *Merchants and Marvels: Commerce, Science, and Art in Early Modern Europe*, ed. Pamela H. Smith and Paula Findlen (New York: Routledge, 2002); and her doctoral dissertation, "Cosmographers vs. Pilots: Navigation, Cosmography, and the State in Early Modern Spain" (University of Wisconsin–Madison, 2001). The reader may also wish to consult an article of mine, "Navigation Techniques and Practices," in *The History of Cartography*, vol. 3, *Cartography in the European Renaissance and Reformation*, ed. David Woodward (Chicago: University of Chicago Press, expected 2005), in which I attempted to focus on the mechanical details of early modern navigational practice and on how many of the new mathematical technologies were actually employed at sea, a subject still much debated.

On early modern English geographical knowledge, one must begin with E. G. R. Taylor's companion works, *Tudor Geography, 1485–1583* (London: Methuen, 1930) and *Late Tudor and Early Stuart Geography, 1583–1650* (London: Methuen, 1934), together with an excellent recent work, Lesley B. Cormack, *Charting an Empire: Geography at the English Universities, 1580–1620* (Chicago: University of Chicago Press, 1997). In describing how and why the English first developed their overseas trading networks, two books were especially useful: David B. Quinn and A. N. Ryan, *England's Sea Empire, 1550–1642* (London: George Allen and Unwin, 1983); and Kenneth R. Andrews, *Trade, Plunder, and Settlement: Maritime Enterprise and the Genesis of the British Empire, 1480–1630* (Cambridge: Cambridge University Press, 1984). Regarding the various techniques and uses of sixteenth-century English cartography, see P. D. A. Harvey, *Maps in Tudor England* (London: Public Record Office and British Library, 1993); and Peter Barber's two companion essays, "England I: Pageantry, Defense, and Government: Maps at Court" and "England II: Monarchs, Ministers, and Maps, 1550–1625," in *Monarchs, Ministers, and Maps: The Emergence of Cartography as a Tool of Government in Early Modern Europe*, ed. David Buisseret (Chicago: University of Chicago Press, 1992). For two excellent articles on the production and use of medieval and early modern portolan charts, see Tony Campbell, "Portolan Charts from the Late Thirteenth Century to 1500," in *The History of Cartography*, vol. 1, *Cartography in Prehistoric, Ancient, and Medieval Europe and the Mediterranean*, ed. J. B. Harley and David Woodward (Chicago: University of Chicago Press, 1987), 371–463; and Corradino Astengo, "The Chart Tradition in the Mediterranean," in *The History of Cartography*, vol. 3, *Cartography in the European Renaissance and Reformation*, ed. David Woodward (Chicago: University of Chicago Press, expected 2005).

The literature devoted to Francis Bacon's life, natural philosophy, and legal and political career is vast. In addressing his philosophical emphasis on practical knowledge, some of the most useful texts are Paolo Rossi, *Philosophy, Technology, and the Arts in the Early Modern Era*, trans. Salvator Attanasio, ed. Benjamin Nelson (New York: Harper and Row, 1970); Paolo Rossi, *Francis Bacon: From Magic to Science*, trans. Sacha Rabinovitch (Chicago: University of Chicago Press, 1968); Antonio Pérez-Ramos, *Francis Bacon's Idea of Science and the Maker's Knowledge Tradition* (Oxford: Clarendon Press, 1988); and Brian Vickers, "Francis Bacon and the Progress of Knowledge," *Journal of the History of Ideas* 53 (1992): 495–518. Though it has been supplanted by these works, one should still consult Benjamin Farrington, *Francis Bacon: Philosopher of Industrial Science* (New York: H. Schuman, 1949). There has been a series of recent works on the connections between Bacon's philosophy and his political career; the most important of these is Julian Martin, *Francis Bacon, the State, and the Reform of Natural Philosophy* (Cambridge: Cambridge University Press, 1992). Stephen Gaukroger, *Francis Bacon and the Transformation of Early-Modern Philosophy* (Cambridge: Cambridge University Press, 2001), addresses many of the same themes as Martin's book, but from a more philosophical perspective. John E. Leary Jr., *Francis Bacon and the Politics of Science* (Ames: Iowa State University Press, 1994), should also be consulted for its close reading of many of Bacon's works. In addition to these books, *The Cambridge Companion to Bacon*, ed. Markku Peltonen (Cambridge: Cambridge University Press, 1996), contains a number of very helpful essays, especially those by Antonio Pérez-Ramos, Paolo Rossi, and Rose-Mary Sargent. Finally, although it does not deal directly with Bacon, Deborah E. Harkness's essay "'Strange' Ideas and 'English' Knowledge: Natural Science Exchange in Elizabethan London," in *Merchants and Marvels: Commerce, Science, and Art in Early Modern Europe*, ed. Pamela H. Smith and Paula Findlen (New York: Routledge, 2002), 137–160, provides a stimulating portrayal of the intellectual, artisanal, and patronage environment in which Bacon lived, worked, and attempted to gain advancement.

The concept of the mathematical practitioner, as set out in E. G. R. Taylor, *The Mathematical Practitioners of Tudor and Stuart England* (Cambridge: Published for the Institute of Navigation at the University Press, 1954), has been a fruitful one in the history of early modern science and technology. This character, or at least the type of practical mathematical knowledge he is supposed to have embodied, has appeared in books and articles examining all parts of western Europe, with a particular emphasis on Italy. For the Continental interpretation of the mathematical practitioner, see Mario Biagioli, "The Social Status of Italian Mathematicians, 1450–1600," *History of Science* 27 (1989): 41–95; M. Henninger-Voss, "Working Machines and Noble Mechanics: Guidobaldo del Monte and the Translation of Knowledge," *Isis* 91 (2000): 233–259; Thomas B. Settle, "The Tartaglia-Ricci Problem: Towards a Study of the Technical Professional in the Sixteenth Century," in *Cultura, scienze e tecniche nella Venezia del cinquecento: Atti del convegno internazionale di studio Giovan Battista Benedetti e il suo tempo* (Venice: Istituto Veneto di Scienze, Lettere ed Arti, 1987); Anthony Grafton, *Leon Battista Alberti: Master Builder of the Italian Renaissance* (New York: Hill and Wang, 2000); Robert S. Westman, "The Astronomer's Role in the Sixteenth Century: A Preliminary Study," *History of Science* 17 (1980): 105–147; and Bruce T. Moran, "German Prince-Practitioners: Aspects in the Development of Courtly Science, Technology, and Procedures in the Renaissance," *Technology and Culture* 22 (1981): 253–274.

Not surprisingly, however, the mathematical practitioner has been an especially well-explored topic in English historiography, where the term originated. After Taylor, the works of J. A. Bennett are vital. They are most valuable for their careful attention to mathematical instruments and their uses, both practical and rhetorical; see his articles "The Mechanics' Philosophy and the Mechanical Philosophy," *History of Science* 24 (1986): 1–28; "The Challenge of Practical Mathematics," in *Science, Culture, and Popular Belief in Renaissance Europe*, ed. Stephen Pumfrey, Paolo L. Rossi, and Maurice Slawinski (Manchester: Manchester University Press, 1991); and "Geometry and Surveying in Early 17th-Century England," *Annals of Science* 48 (1991): 345–354. Stephen Johnston has written a benchmark article on the subject, "Mathematical Practitioners and Instruments in Elizabethan England," *Annals of Science* 48 (1991): 319–344, as well as a most useful and engaging doctoral dissertation, "Making Mathematical Practice: Gentlemen, Practitioners and Artisans in Elizabethan England." Both works do much to flesh out the mathematical practitioner by means of more detailed career biographies than Taylor was able to provide in her comprehensive book.

Several authors have addressed the patronage and employment of English mathematical practitioners and the uses to which their knowledge was put by their social superiors; see in particular Frances Willmoth, *Sir Jonas Moore: Practical Mathematics and Restoration Science* (Woodbridge, Suffolk: Boydell Press, 1993); Frances Willmoth, "Mathematical Sciences and Military Technology: The Ordinance Office in the Reign of Charles II," in *Renaissance and Revolution: Humanists, Scholars, Craftsmen, and Natural Philosophers in Early Modern Europe*, ed. J. V. Field and Frank A. J. L. James (Cambridge: Cambridge University Press, 1993); A. J. Turner, "Mathematical Instruments and the Education of Gentlemen," *Annals of Science* 30 (1973): 51–88; and Cormack, *Charting an Empire*. Other authors have sought to complicate both the image and the context of the mathematical practitioner. Katherine Neal, "Mathematics and Empire, Navigation and Exploration: Henry Briggs and the Northwest Passage Voyages of 1631," *Isis* 93 (2002): 435–453, has shown that the distinction between mathematical and nonmathematical navigators could sometimes be rather vague. Keith Thomas, "Numeracy in Early Modern England," *Transactions of the Royal Historical Society* 37 (1987): 103–132, provokes important questions concerning the frequency and depth of early modern mathematical knowledge. Mordechai Feingold, *The Mathematicians' Apprenticeship: Science, Universities, and Society in England, 1560–1640* (Cambridge: Cambridge University Press, 1984), demonstrates convincingly that the English universities were neither as antimathematical nor as antipractical as they have often been portrayed.

Some of the original seeds for this book were sown in a graduate seminar at Princeton University, cotaught by Professors Michael Mahoney and Mary Henninger-Voss, which examined the history of practical knowledge and the engineer in early modern Europe. Among the readings assigned in the seminar were Nicholas Adams, "Architecture for Fish: The Sienese Dam on the Bruna River—Structures and Designs, 1468–ca. 1530," *Technology and Culture* 25 (1984): 768–797; Hélène Vérin, *La gloire des ingénieurs: L'intelligence technique du XVIe au XVIIIe siècle* (Paris: Albin Michel, 1993); and a manuscript by Professor Mahoney, "Organizing Expertise: Engineering and Public Works under Colbert, 1662–83." All three of these pieces, and especially the Adams article, made me consider how early modern people convinced one another that they knew how to do something, in the absence of conventional credentials—a key theme of this book. I also came to be very interested in the patronage strategies of early modern rulers who wished to control and maintain their access to particular intellectual talents and technical skills. On this theme, the most helpful works for me have been Lisa Jardine and Anthony Grafton, "'Studied for Action': How Gabriel Harvey Read His Livy," *Past and Present* 129 (1990): 30–78; William H. Sherman, *John Dee: The Politics of Reading and Writing in the English Renaissance* (Amherst: University of Massachusetts Press, 1995); Pamela H. Smith, *The Business of Alchemy: Science and Culture in the Holy Roman Empire* (Princeton, N.J.: Princeton University Press, 1994); Mario Biagioli, *Galileo, Courtier: The Practice of Science in the Culture of Absolutism* (Chicago: University of Chicago Press, 1993); J. R. Hale, "Tudor Fortifications: The Defence of the Realm, 1485–1558," in *The History of the King's Works*, ed. H. M. Colvin (London: H. M. Stationery Office, 1982), 4:367–401; Lesley B. Cormack, "Twisting the Lion's Tail: Practice and Theory at the Court of Henry Prince of Wales," in *Patronage and Institutions: Science, Technology, and Medicine at the European Court, 1500–1700*, ed. Bruce T. Moran (Rochester, N.Y.: Boydell

Press, 1991); and Pamela O. Long, "Power, Patronage and the Authorship of *Ars:* From Mechanical Know-How to Mechanical Knowledge in the Last Scribal Age," *Isis* 88 (1997): 1–41. As I explored these themes in Elizabethan England, it was both informative and encouraging to note that others had encountered similar trends in other countries; specifically, Henry Heller, *Labour, Science, and Technology in France, 1500–1620* (Cambridge: Cambridge University Press, 1996); and David C. Goodman, *Power and Penury: Government, Technology, and Science in Philip II's Spain* (Cambridge: Cambridge University Press, 1988). On the tension faced by those possessing technical skills, concerning how much to keep secret and how much to advertise one's knowledge, see Pamela O. Long's excellent and comprehensive recent book, *Openness, Secrecy, Authorship: Technical Arts and the Culture of Knowledge from Antiquity to the Renaissance* (Baltimore: Johns Hopkins University Press, 2001).

One of the more contentious debates in the historiography of early modern England concerns the development of an ever more powerful royalist regime, centered on the monarch but administrated by the Privy Council, which itself evolved to become a much more efficient and effective governing body. This change is usually seen as originating with the administration of Thomas Cromwell during the reign of Henry VIII and continuing (rather unevenly) through the late Tudor and early Stuart periods, reaching a point of crisis around 1640. The foundation for this argument is the work of G. R. Elton, *The Tudor Revolution in Government: Administrative Changes in the Reign of Henry VIII* (Cambridge: Cambridge University Press, 1953). Other works exploring and developing this idea include Alan G. R. Smith, *The Government of Elizabethan England* (London: Edward Arnold, 1967); Joel Hurstfield, *Freedom, Corruption, and Government in Elizabethan England* (Cambridge, Mass.: Harvard University Press, 1973); Penry Williams, *The Tudor Regime* (Oxford: Clarendon Press, 1979); and David Loades, *Power in Tudor England* (New York: St. Martin's Press, 1997). Winthrop Hudson, *The Cambridge Connection and the Elizabethan Settlement of 1559* (Durham, N.C.: Duke University Press, 1980), traces the origins of the well-educated and confident corps of ministers serving on Elizabeth's Privy Council to the rise of humanist education at Cambridge. The works of Michael A. R. Graves and Patrick Collinson have done much to reveal the Elizabethan Privy Council's growing belief in its own governing authority, as well as its reliance upon trusted clients to carry out its orders, especially in managing the affairs of the House of Commons; see Graves, *Thomas Norton, the Parliament Man* (Oxford: Blackwell, 1994); Collinson, "The Monarchical Republic of Elizabeth I," 394–424, and "The Elizabethan Exclusion Crisis and the Elizabethan Polity," *Proceedings of the British Academy* 84 (1993): 51–92. G. R. Elton, *The Parliament of England, 1559–1581* (Cambridge: Cambridge University Press, 1986); David Dean, *Law-Making and Society in Elizabethan England: The Parliament of England, 1584–1601* (Cambridge: Cambridge University Press, 1996); and Michael A. R. Graves, "Managing Elizabethan Parliaments," in *The Parliaments of Elizabethan England*, ed. D. M. Dean and N. L. Jones (Oxford: Basil Blackwell, 1990), all discuss in detail the Elizabethan Privy Council's strategies for managing Parliament. J. E. Neale's two companion works, *Elizabeth I and Her Parliaments, 1559–1581* (London: Jonathan Cape, 1953) and *Elizabeth I and Her Parliaments, 1584–1601* (London: Jonathan Cape, 1957), even as they portray the House of Commons as an increasingly institutionalized and independent body, nevertheless offer a telling portrayal of the Privy Council's efficacy in controlling the legislative agenda there.

Much recent work has attempted to qualify or undermine Elton's emphasis on the Privy Council, particularly in highlighting the ongoing importance of other elites and governing institutions in early modern England (the court, Parliament, the nobility, the gentry, etc.). These alternate centers of power, and especially the land-owning gentry, are portrayed as the indispensable key to the Privy Council's success in governing at the local level; as local officeholders, they could either facilitate the council's governance through their active cooperation or else undermine it through their resistance. Steve Hindle, *The State and Social Change in Early Modern England, c. 1550–1640* (Basingstoke, Hampshire: Palgrave, 2000); and Michael L. Braddick, *State Formation in Early Modern England, c. 1550–1700* (Cambridge: Cambridge University Press, 2000), have emphasized the extent to which the authority of the Privy Council in the localities depended upon the acquiescence and active cooperation of local elites. A. Hassell Smith, *County and Court: Government and Politics in Norfolk, 1558–1603* (Oxford: Clarendon Press, 1974); and Diarmaid MacCulloch, *Suffolk and the Tudors: Politics and Religion in an English County, 1500–1600* (Oxford: Clarendon Press, 1986), both show that as the Privy Council worked to assert greater control over local affairs, their efforts did not go unchallenged by the gentry.

Divisions between the center and the locality should not be overstated, however. The relationship between the Privy Council and local officials was much more of a dependent partnership than a contentious opposition, and the eventual taking of sides in the English Civil War during the 1640s does not follow any neat pattern with respect to "local" versus "royal" interests. The historians of the revisionist school have undermined the traditional historiographical "court and country" dichotomy for the period leading up to the Civil War, arguing that the distinction would have made little sense to members of the House of Commons, who most often served as royally appointed justices of the peace, lord-lieutenants of the militia, and other agents of the Crown in their own localities. See Conrad Russell, *Parliaments and English Politics, 1621–1629* (Oxford: Clarendon Press, 1979), and *Unrevolutionary England, 1603–1642* (London: Hambledon Press, 1990); J. S. Morrill, *The Revolt of the Provinces: Conservatives and Radicals in the English Civil War, 1630–1650* (New York: Barnes and Noble Books, 1976); Michael Hawkins, "The Government: Its Role and Its Aims," and L. M. Hill, "County Government in Caroline England, 1625–1640," both essays in *The Origins of the English Civil War*, ed. Conrad Russell (London: Macmillan, 1973); and D. M. Dean, "Parliament and Locality," in *The Parliaments of Elizabethan England*.

# Index

Numbers in **bold type** denote illustrations.

active life, 12–16
Agricola, Georgius, 19, 53; career of, 23; *De re metallica*, 23–30
Antoniszoon, Cornelis, *The Safeguard of Sailers* (trans. Norman), **95**
Antwerp, wool and cloth market of, 97
apprenticeships: in copper battery, 38; of ships' masters and pilots, 92, 120, 137
Arctic exploration: English, 87–89; particular difficulties of, 110, 115. *See also* exploration, voyages of
Aristotle, 188, 192–93
*Arte of Nauigation, The* (Cortés/Eden), 87, 98, 117–18, **119**, 142–45, 200; latitude calculation, 143–45
artisans. *See* practitioners
Ascham, Roger, *The Scholemaster*, 180
Ashley, Anthony, translation of *The Mariners Mirrovr*, **100**
astrolabe: mariner's, 99, **100**, **102**; universal, 99, 136. *See also* navigational instruments
astronomical instruments, 89. *See also* navigational instruments; navigational technologies
Augsburg, center of mining expertise, 19

Bacon, Elizabeth, 176–77, 179, 243n. 67
Bacon, Francis, 18; early career of, 192–99; humanism and, 193–94, 203; legacy of, 186–87; letter to Burghley, 194–95; member of Parliament, 194; quest for patronage, 192, 194–99, 203, 211–12; reform of common law, 205; reform of natural philosophy, 187–92; role as expert mediator, 203–4, 212; 16th-century context of, 187, 199–200
Bacon, Francis, works of: *Great Instauration*, 192, 195, 199; "In Praise of Knowledge,"

196; *New Atlantis*, 192, 197, 207–11; *New Organon*, 186, 204–5; *Of the Advancement of Learning*, 197–99, 207–8; "Of the Interpretation of Nature," 205–6; "Parasceve," 198, 206–7; speeches for *Gesta Grayorum*, 197
Bacon, Nicholas, 193
Baconian natural philosophy: vs. alchemical tradition, 191, 196; black-boxing and, 204, 210–11; direct observation of nature, 188, 201–2; early formulation of, 195–99; elitism of, 205–11; expertise and, 199–212; hierarchy within, 205–6, 207, 210; importance of proper method, 190–91, 202–3; inclusiveness of, 191, 204–5; "leveling of wits," 191, 204–5; mechanical arts and, 188–90, 202; need for collaboration, 190–91, 200–201, 205; practical utility and, 189–90, 200, 201–2; role of artisans in, 190–91; vs. Scholasticism, 187–88, 189, 192–93, 196; secrecy within, 205–6, 209, 211; in service of state, 207
Baker, Matthew, 69, 71, 72, 73, 106–7
*Barke Aucher*, 106–7
Barrey, Richard, 61, 68, 73
Bedwell, Thomas, 67; career of, 230n. 33
Bensalem. *See New Atlantis*
Best, George, 128, 132
Biringuccio, Vannoccio, 221n. 8
black-boxing, 150; Baconian natural philosophy and, 204, 210–11; navigation manuals and, 149–52, 164–65, 167–68, 172–73, 176, 182
Blundeville, Thomas: career of, 176–77; lack of nautical experience, 178; *M. Blundevile, His Exercises . . .* , 176–79, 200, 202
Bodenham, Roger, 106–7
Bodin, Jean, 248n. 53

"Boke of Idrography" (Rotz), **104–5**
book learning: as basis for expertise, 11; humanism and, 15–16; vs. practical experience, 13
Borough, Stephen: Arctic exploration and, 11, 113–17; *The Arte of Nauigation* and, 117–18, 123; Richard Chancellor and, 113; guest at Casa de Contratación, 117, 118, 133; petition to become chief pilot of England, 120–22, 201; recording navigational data, 114, 116–17; role as expert mediator, 91, 116–17, 132–33; Russian pilots and, 114–15; source of his authority, 91, 133; training of English pilots, 119–24, 135; use of mathematical navigation, 113–14, 116; voyage to River Ob', 113–17
Borough, William, 200; assistance to Martin Frobisher, 125–26; career of, 154; consultations on Dover Harbor, 60, 61, 62, 64, 66, 69, 71, 72–73; *A Discovrs of the Variation of the Cumpas*, 135, 153–57, 203; instructions to Jackman and Pett, 130–31; on nautical experience, 183, 202; opinion of John Trew, 63–64; pilot for Muscovy Company, 126, 154
Bourne, William: career of, 143; *A Regiment for the Sea*, 87, 132, 142–53, 162, 200, 203; role as expert mediator, 153
Boys, Edward, 70, 75–76
brass cannon. *See* naval ordnance
brass wire, 37
*Breve compendio . . .* (Cortés), 117
Burghley, William Lord. *See* Cecil, William

Cabot, John, 98, 106; voyages of exploration, 97, 107
Cabot, Sebastian: Stephen Borough and, 113–14; Casa de Contratación and, 106, 110, 133; instructions of, 107–11; Muscovy Company and, 107; return to English service, 106; role as expert mediator, 91, 132–33; source of his authority, 91, 133; training of English mariners, 106–7, 135, 200
Cambridge University, book learning and, 13
cannon. *See* naval ordnance
Casa de Contratación (Seville), 91, 103–6, 110, 118, 120–21, 235n. 30; Stephen Borough and, 117, 118, 133; Sebastian Cabot and, 106, 110, 133
Cecil, William, 19; Francis Bacon and, 194–95, 197; Company of Mines Royal and, 34, 41, 51–52; Dover Harbor and, 72, 78; Lionel Duckett and, 51; Martin Frobisher and, 124–25; William Humfrey and, 19, 35–36, 37–39, 50, 53; introduction to court, 220n. 26; Johannes Loner and, 41, 51–52; George Nedham and, 51; Thomas Thurland and, 20–21, 44–45, 46–48
centralization of royal government: in England, 7–8, 82–86, 214–15; in western Europe, 7–8, 214–15. *See also* Privy Council
*Certaine Errors in Navigation* (Wright), 135, 202, 203; audience for, 165–67; black-boxing and, 167–68; compared with Thomas Harriot's "Instructions," 169, 172–73; mathematical sophistication of, 166–69; Mercator projection charts, 167–69; use of tables, 167, 168–69
Chancellor, Richard, 107; Stephen Borough and, 113; death of, 112, 116; pilot major of Muscovy Company, 111–13
charts. *See* navigational charts
Cheke, John, 220n. 26
circumpolar charts. *See* navigational charts
civilian militia (of England), 2, 241n. 34; "trained bands" of London, 158–61
Clark, John, 58–59
cloth trade (English): decline of, 97; in the Low Countries, 88, 91; in Spain, 2
coastal piloting. *See* piloting
Cobham, William Lord, 60–61, 63
colonization. *See* North America
Company of Mineral and Battery Works, 39, 49–50
Company of Mines Royal, 10–11, 17, 19; account books of, 51, 222n. 20; cooperation within, 45–49; expenses of, 39–41, 50, 52; mistrust and suspicion within, 19–22, 30, 43–45, 53–54; Privy Council support for, 31, 34; profits of, 52–53; Scottish gold and, 46–48. *See also* English shareholders; German shareholders
compass. *See* magnetic compass
consensus. *See* Dover Harbor, rebuilding of; expert mediators
contemplative life, 14–15
copper: export of, 37, 49, 53; manufactured goods, 36–37, 49–50; marketing of, 36, 39, 49–50, 52, 53

copper battery, 37–39. *See also* Company of Mineral and Battery Works; Humfrey, William; Schutz, Christopher

copper manufacturing expertise: English need for, 38; German control of, 36–37. *See also* copper; copper battery

copper mines: English, 4, 6, 17, 19, 34, 49, 53; German, 23

copper ore: assaying of, 32, 34–36; extraction of, 24; prospecting for, 23–24; smelting of, 24–28, 223n. 36. See also *De re metallica*

Cortés, Martín: *The Arte of Nauigation* (trans. Eden), 87, 98, 117–18, **119**, 142–45, 200; *Breve compendio* . . . , 117

Courteney, William, 4–5

craft knowledge. *See* practical knowledge

craft practitioners, craftsmen. *See* practitioners

craft secrecy. *See* secrecy

Crawford Moor (Scotland): search for gold at, 45–49

Cromwell, Thomas, 59, 83

cross-staff, 99, **100**, **101**, 135–36; difficulty in using, 167, 172–73. *See also* navigational instruments

Cumberland (Cumbria), copper mines in, 6, 19, 34, 49, 53

Davis, John: Arctic explorations of, 88; career of, 163; *The Seamans Secrets*, 163–65, 202

dead reckoning, 93–94, 112

Dee, John, 10, 61, 138, 200; circumpolar charts of, 127, 136, 238n. 68; "Mathematicall Praeface," 135–36; navigational instruction of, 111–12, 125–28; patronage of, 247n. 29

delegation of royal authority, 7–8

*De re metallica* (Agricola): illustrations, **25–27**, **29**; mining administration and management, 28–30; mining machinery, 24–27; ore extraction, 24; processing and smelting ore, 24, 28; prospecting for ore, 23–24

*De tradendis disciplinis* (Vives), 12

De Vos, Cornelius, 45–48; secrecy of, 46

Digges, Leonard, 61

Digges, Thomas, 55, 64, 183; career of, 61; disinterestedness of, 72; Dover commission and, 61, 84; harbor mouth and, 76–77; lack of nautical experience, 183, 202; master surveyor at Dover Harbor, 66–67, 71–72;

mathematical works of, 61; plan to rebuild Dover Harbor, 66; political connections of, 61, 71, 84, 230n. 34, 231n. 52; role as expert mediator, 56–57, 67, 72, 74, 78–79, 84–85, 141–42, 215; Romney method and, 70–72, 74

*Discours admirables* (Palissy), 193

*Discovrs of the Variation of the Cumpas, A* (W. Borough), 135, 203; audience for, 153–55, 157; mathematical sophistication of, 154–57

Dour River, 57, 62

Dover, Town of: growth of, 79; local authorities, 56; location of, 57; mayor of, 61, 68, 76; petitions for relief, 59, 60; Spanish Armada and, 4–5

Dover commission: formation of, 60–61, 83; harbor mouth and, 74, 76–77; overruled by Privy Council, 62–63, 66, 68, 74, 83–84; Romney method and, 70–71, 84; John Trew and, 63–66, 83

Dover Harbor, 4–5: early renovations of, 56, 58–59, 82–83; geographic circumstances of, 57–58; importance of, 55; plans of, 64–**65**, **80–81**; silting up of, 57–59, 62, 79; surveys of, 59–60, 62, 66. *See also* Dover Harbor, rebuilding of

Dover Harbor, rebuilding of, 17; backwater pent, 62, 79; centralized administration of project, 56–57; completion of, 79; construction of seawalls, 57, 62, 69–74; cost of, 60, 62, 63, 64–66, 67, 70, 74, 77; Dover mariners and, 59, 76, 77; Flemish proposals for, 62, 66; importance of consensus in, 57, 74, 77–79, 85; location of harbor mouth, 57, 74–78; Romney method and, 68–74, 84; royal funding for, 58–59, 62, 83–84; sluice gates, 62, 73

Drake, Francis, 5, 64

Duckett, Lionel, 41, 48, 51

Dudley, John. *See* Northumberland, Duke of

economic nationalism, 43

Eden, Richard: on Stephen Borough, 11, 118; on Sebastian Cabot, 107; translation of *The Arte of Nauigation*, 117–18, 119

education: for the active life, 12–15; humanism in England, 12–16. *See also* humanism

Edward VI (king of England), 106

Elizabeth I (queen of England), 1, 31, 33–34, 47–48, 49, 119–20, 195–96
English Channel: currents and tides of, 57–58; difficulty in piloting, 93
English coins, refining of, 20, 35
English harbors, ruin of, 58, 227n. 4, 228n. 9
English navigation, advances in, 3, 88–92. *See also* navigation
English shareholders (in Company of Mines Royal), 21, 33–34; in default of payments, 39–42, 50–52; petition to Queen Elizabeth, 32–33; suspicions of, 19, 32–33, 39–42, 44, 51–52; vulnerability of, 31–32, 38, 43–45, 53–54
"Erection of an Achademy in London, The" (Gilbert), 1, 13–15, 180–81
experience: as basis for expertise, 10, 213; vs. skill, 10–11, 213; vs. theoretical understanding, 8. *See also* nautical experience
experimentation, 14
expert, definition of, 10–11, 218n. 13
expertise: basis for, 10–12, 16–18, 56, 82, 84–86, 133–34, 213; changing definition of, 11, 16–18, 82, 133–34, 213; empirical vs. theoretical, 10–11; management of, 17, 53–54; mathematics and, 139–42, 179–82, 214–15; social status and, 17–18, 134; as source of authority, 22, 54, 85. *See also* Baconian natural philosophy; hydraulic construction; mathematics; mining expertise; navigation
expert mediators: Baconian natural philosophy and, 210–11; building consensus, 57, 74, 78–79, 85; Company of Mines Royal and, 22, 54; disinterestedness of, 71–72; Dover Harbor and, 56, 64–66, 84–86; issues of trust, 21–22, 53–54, 82–83; as knowledge brokers, 8, 213–16; mathematics and, 139–42, 179–82, 214–15; navigation and, 91, 132–33, 138–39; need for, 8–9, 16, 64–66, 78–79, 85, 214–16; patronage of, 8–9, 11, 17, 138–40, 214; vs. practitioners, 213–14; theoretical knowledge and, 16; tool of centralized management, 8–9, 17, 53–54, 84–85, 138–40, 214–16. *See also* Bacon, Francis; Borough, Stephen; Cabot, Sebastian; Digges, Thomas; mathematicians; Thurland, Thomas
exploration, voyages of: English investment in, 98; Iberian, 98–99; Muscovy Company and, 107–13; preparation for, 124; records of, 92, 124; vs. routine navigation, 92, 98, 117, 238n. 68. *See also* Borough, Stephen; Cabot, John; Davis, John; Frobisher, Martin; Jackman, Charles; Pett, Arthur; Willoughby, Hugh

Fenlands, reclamation of, 215–16
fortification, 14, 138, 181
Frobisher, Martin, 10; competence of, 128; ignorance of mathematical navigation, 126–28; nautical experience of, 124, 126; receiving navigational instruction, 125–28; search for northwest passage, 124–28

*Gargantua* (Rabelais), 13
Gaukroger, Stephen, 194
German mine workers: Company of Mines Royal and, 34; competence and integrity of, 35, 53; marriages to English women, 224n. 55; taking offense, 21, 43, 44; violence toward, 42–43; wages in arrears, 41
German shareholders (in Company of Mines Royal), 19, 31, 33; resentment of, 40–42, 44, 50–52, 54. *See also* Haug, Langnauer, and Company
*Gesta Grayorum*. *See* Gray's Inn
Gilbert, Humphrey: career of, 13; "The Erection of an Achademy in London," 1, 13–15, 180–81
gold, search for in Scotland, 45–49
Gray's Inn, 192, 194; *Gesta Grayorum*, 196–97; Privy Council and, 197
*Great Instauration* (Bacon), 192, 195, 199
Gresham, Thomas, 44
Gresham College, 13
gun foundries, English, 4, 6

Hakluyt, Richard, 64, 92; mathematical lectureship and, 157–58; *Principall Navigations*, 87, 92, 113
Hall, Christopher, 125, 128, 237n. 67, 238n. 72
Harriot, Thomas: "Arcticon," 169; career of, 169; "Instructions for Raleigh's voyage to Guyana," 169, 172–76; tutor to Walter Raleigh, 169
Haug, Langnauer, and Company, 19, 31, 32, 48, 50, 52

Hawkins, John: consultations on Dover Harbor, 77–78; naval philosophy of, 2–5
Hechstetter, Daniel, 20–21; assay of English ore, 32, 34–36; career of, 32; complaints of, 41, 42–44; expertise of, 10–11, 31; leasing of royal mining rights, 52; royal patentee, 33–34; Scottish gold and, 47–48; secrecy of, 38–39
Hechstetter, Joachim, 46
Henry VII (king of England), 58, 97, 106
Henry VIII (king of England), 6, 59, 82–83, 97, 98, 104, 106, 138
Hood, Thomas: audience for printed works of, 158–63; career of, 157–58; *The Marriners guide*, 159–62, 201; mathematical lecturer, 158–61, 201, 241n. 33; private tutorials of, 162–63; use of dialogue style, 161–62
Hooke, Robert, *Micrographia*, 187
humanism: Francis Bacon and, 193–94, 203; book learning and, 15–16; at English universities, 12, 180, 193; mathematics and, 180–81; practical knowledge and, 12–13, 193–94, 199, 202, 203, 213; role in shaping expertise, 12–16, 213–14; technical treatises and, 193
Humfrey, William, 19, 53; botched assay of, 34–36, 43, 54; career of, 34–35; manufacture of copper goods, 37–39, 49–50; mistrust of German mine managers, 36–39, 43
hundredweight, 223n. 34
hydraulic construction, 60; Dutch/Flemish methods, 61–62, 66, 67, 69, 84; expertise in, 56, 63, 82

Iberian nautical advances, 3
Inns of Court, 220n. 27
"In Praise of Knowledge" (Bacon), 196
instructional manuals: publication of, 15–16, 133, 140, 214; use of vernacular, 140. *See also* navigation manuals
"Instructions for Raleigh's voyage to Guyana" (Harriot): audience for, 169, 172; black-boxing and, 172–73, 176; clarity of, 172–74; compared with Wright's *Certaine Errors in Navigation*, 169, 172–73; latitude calculation, 173–76; use of tables, 172–73
investors. *See* Company of Mines Royal; merchants

iron cannon. *See* naval ordnance
Ivan IV (czar of Russia), 112

Jackman, Charles: career of, 129; receiving navigational instruction, 129–31; recording navigational data, 130–31; search for northeast passage, 129–31
James I (king of England), 198
Jenkinson, Anthony, 235n. 28
Johnston, Stephen, 179
joint-stock companies: formation of, 9, 107, 108; German mines and, 30. *See also* Company of Mineral and Battery Works; Company of Mines Royal; Muscovy Company
justices of the peace, 7, 14

Keswick: Company of Mines Royal headquarters, 20, 34; hostility toward German miners, 42–43, 44
Kishlansky, Mark A., 78
*Kunstkammern*, 189

latitude, calculation of, 89, 99–103, 130, 142, 240n. 16; William Blundeville and, 178; William Bourne and, 145–49, 241n. 21; Martín Cortés and, 143–45; John Davis and, 164; Thomas Harriot and, 173–76
latitude observations: by Stephen Borough, 114, 116; by Hugh Willoughby, 112
lead and line, 94–95, 99, **100**
Leary, John E., 205
Leicester, Earl of: Company of Mines Royal and, 40; Thomas Digges and, 61
literacy, of English mariners, 109
local knowledge, 213, 215; vs. mathematical navigation, 103, 131–34; navigation and, 89–90, 92–97, 98
Lok, Michael, 125–27
Loner, Johannes, 19, 20–21; complaints of, 41, 50–52; founding of Company of Mines Royal and, 31; refining English coins, 35; secrecy of, 39
Long, Pamela, 15
longitude, 89
lords lieutenant of English militia, 7. *See also* civilian militia
Low Countries, water management in, 8. *See also* hydraulic construction

magnetic compass, 93, **100**; problems with, 94
magnetic variation, 94, 110–11, 115, 130, 236n. 32; Stephen Borough and, 114, 116; William Borough and, 154; compass of variation, **115**; Martin Frobisher and, 128; Edward Wright and, 167
management: crises of, 17, 22, 53–54, 68; of technical projects, 8–9, 53–54
manual arts. *See* practical knowledge
mariners, English, 3. *See also* Dover Harbor, rebuilding of; pilots
*Mariners Mirrovr, The* (Wagenaer/Ashley), **100**
*Marriners guide, The* (Hood), 159–62, 201
Martin, Julian, 205
Mary (queen of England), 112, 117
Mary (queen of Scotland), 48
mathematical arts, vs. mathematical practitioners, 185, 244n. 84
mathematical lectureship (London): audience for, 158–61; foundation of, 157–58; Thomas Hood and, 158–61, 201, 241n. 33; Edward Wright and, 165
mathematical navigation, 103, 131–34, 185; vs. maritime practice, 136; mastery of, 135–36; perceived value of, 90. *See also* mathematics; navigation; navigation manuals
mathematical practitioners, critical reexamination of, 140–41, 183–85
"Mathematicall Praeface" (Dee), 135–36
mathematicians: vs. craft practitioners, 181–83; divisiveness of, 141–42, 181–82; growing authority of, 136–40; hostility toward, 179–80; lack of nautical experience, 136–37; offering navigational instruction, 111–12, 125–28, 130; patronage of, 138–40; role as expert mediators, 139–42, 179–82, 214–15
mathematics: expertise and, 136, 139–42, 179–82, 214–15; practical uses of, 10, 14, 135–36, 179–81; theoretical knowledge and, 90, 214; use in navigation, 17, 89–91, 98–103, 131–34, 135–36, 185; use in surveying, **29**–30, 138, 179–80
*M. Blundevile, His Exercises . . .* (Blundeville), 200; audience for, 177–79; contents of, 177–78; latitude calculation, 178; organization of, 177–78, 200, 202
mediators. *See* expert mediators
Medina Sidonia, Duke of, 2, 6

Mercator projection charts. *See* navigational charts
merchant marine (English): expansion of, 6–7; involvement in naval combat, 4
merchants (English): cooperation among, 107, 108; need for expert mediators, 9; promotion of mathematical navigation, 18, 91, 103–6, 131, 139, 159, 162–63
*Micrographia* (Hooke), 187
militia. *See* civilian militia
mine management, 28–30; English managers, 20, 42–44; German managers, 20, 31
mining, early: English (tin, lead, iron), 22; German (copper), 23
mining expertise: English lack of, 20, 31, 35–36, 53–54; English need for, 21–22, 33, 44–45, 51, 53–54; German control of, 19, 23–30, 31, 46, 48–49
mining law, 28–30
mining machinery, 23, 24, **25-27**
mining treatises, 23. *See also* Agricola, Georgius; *De re metallica*
More, Thomas, *Utopia*, 12–13
Muscovy Company: Stephen Borough and, 117, 123; commission of Jackman and Pett, 129–131; Martin Frobisher and, 124–25; maiden voyage of (Willoughby), 107–13; organization of, 107, 108; Russia trade of, 112, 116

natural philosophy. *See* Baconian natural philosophy
nautical experience: basis for expertise, 91; English need for, 106–7, 110; hierarchy of ranks at sea, 120, 122; mathematicians' lack of, 136–37; navigational knowledge and, 95–97, 110
naval engagements, English strategy for, 3
naval ordnance: brass cannon, 4, 36, 49; iron cannon, 4
naval shipyards, 6
navigation, 14, 181; advances in, 3, 98–99; expertise in, 91; mathematization of, 17–18, 89–91, 98–103, 113, 131–34, 185; teaching of, 14, 106–11, 117–24. *See also* navigation manuals; navigational technologies
navigational charts, 89, 94, 159, **160**; circumpolar, 127, 130, 136, 238n. 68; latitude scales of, 103, **104-5**; Mercator projection,

167–69, **170–71**, 242nn. 56–57. *See also* navigational technologies
navigational data: recording of, 109–10, 114, 116–17, 130–31
navigational instruments, 3, 14, 17, 99, **100– 102**, 127; construction of, 118–19; included within navigation manuals, 118, **119**, 150– 51, **152**, 164; provided for English pilots, 111, 126–27. *See also* navigational technologies
navigational tables: Thomas Harriot and, 172–73; Edward Wright and, 167. *See also* navigational technologies
navigational technologies, 98–103; English ignorance of, 89–91, 109–11, 120–21; in Iberia, 98–99, 106, 127; introduction of in England, 89–92, 109–11, 118, 132–33
navigation manuals, 18, 124; audience for, 137– 39; by William Blundeville, 176–79; by William Borough, 153–57; by William Bourne, 142–53; compared with Baconian natural philosophy, 199–204; by Martín Cortés, 117–19, 142–45; by John Davis, 163–65; by Thomas Harriot, 169, 172–76; by Thomas Hood, 157–63; sophistication of, 136, 140–41; by Edward Wright, 165–69
navigators. *See* pilots
Nedham, George, 48, 51
*New Atlantis* (Bacon), 192, 197; elitism and, 208–11; experimental facilities in, 209; expert mediation and, 210–11; Fathers of Salomon's House, 208–11; hierarchy in, 210; Salomon's House, 207–11; secrecy and, 209, 211
*New Organon* (Bacon), 186, 204–5
*Newe Attractiue . . . , The* (Norman), 154, 178, 182
Norman, Robert, 183; *The Newe Attractiue . . .*, 154, 178, 182; translation of *The Safeguard of Sailers*, **95**
North America, colonization of, 13, 89, 169
northeast passage to Asia, search for, 87, 89, 107, 129. *See also* Borough, Stephen; Jackman, Charles; Pett, Arthur; Willoughby, Hugh
North Sea, piloting in, 89, 94–95
Northumberland, Duke of (John Dudley), 98, 106
Northumberland, Earl of (Thomas Percy), 49

northwest passage to Asia, search for, 13, 89, 124–25. *See also* Davis, John; Frobisher, Martin
Nuremburg, center of copper manufacture, 38

obligatory point of passage, 17, 140, 181–82, 215; black-boxing and, 150, 210–11
*Of the Advancement of Learning* (Bacon), 197– 99, 207–8
"Of the Interpretation of Nature" (Bacon), 205–6
overseas trade. *See* trade, overseas
Oxford University: book learning and, 13; mathematics and, 179

Palissy, Bernard, *Discours admirables*, 193
Palmer, Henry, 75–76
Paradise (Dover Harbor), 58, 59, 73
"Parasceve" (Bacon), 198, 206–7
Parliament, selection of members, 8
Parma, Duke of, 2, 6
patents. *See* royal patents
patronage: of Baconian natural philosophy, 199; of mathematicians, 138–42, 162–63; of practical knowledge, 15–16, 138–40; role in shaping expertise, 11–12, 16–17
pent. *See* Dover Harbor, rebuilding of
Pérez-Ramos, Antonio, 186
Pett, Arthur: career of, 129; receiving navigational instruction, 129–31; recording navigational data, 130–31; search for northeast passage, 129–31
Pett, Peter, 69, 71, 72, 73, 75
Philip II (king of Spain), 1, 3, 117
piloting: difficulty of following the coast, 93; traditional art of, 92–94. *See also* dead reckoning
pilots (English), 3; cooperation among, 108– 10, 129, 200–201; lack of education, 94, 109; lack of experience, 108–10; maritime practices of, 91–92, 136, 182, 185; mathematical deficiencies of, 94, 153, 154–55, 163, 166, 182; navigation manuals and, 137–39; training of, 106–7, 119–24
plane charts. *See* navigational charts
pole star, and latitude measurement, 99
portolan charts. *See* navigational charts
Portuguese navigation, 3; exploration and, 98–99. *See also* navigation

Poyntz, Fernando, 69, 70, 72–73, 74–75, 83; chief overseer at Dover Harbor, 66–68; departure of, 73
practical knowledge, 11–12, 138; humanism and, 12–13, 193–94, 202, 203, 213–14; secrecy and, 15–16
practitioners: vs. experts, 11–12, 182, 213–14; lack of theoretical knowledge, 16–17, 182; vs. mathematicians, 181–83; social status of, 182–83
*Principall Navigations* (Hakluyt), 87, 92, 113
Privy Council: Stephen Borough and, 120–22; Company of Mines Royal and, 31, 34; Thomas Digges and, 61, 67, 72, 76–77; Dover Harbor and, 55–56, 59–60, 62–68, 71–74, 76, 78–79, 82–86; efforts toward centralization, 7–8, 214–15, 218nn. 9–10; England's defense and, 3–4; *Gesta Grayorum* and, 197; need for expert mediators, 7–8, 64–66, 214–15; support for English navigation, 106, 119–22; support for joint-stock companies, 9, 107

quadrant, 99, **100**

Rabelais, François, *Gargantua*, 13
race-built English galleons, 3
Radcliffe, Lady, 42
Raleigh, Walter, 169
Ramus, Peter, 174, 193
Rawley, William, 192, 211
Recalde, Juan Martínez de, 217n. 5
Recorde, Robert, 111, 127, 155
refining coins. *See* English coins
*Regiment for the Sea, A* (Bourne), 87, 132, 200, 203; audience for, 143, 153; black-boxing and, 149–52; compared with *The Arte of Nauigation*, 143–49; latitude calculation, 145–49, 178, 241n. 21; mathematical explanation in, 150–53; popularity of, 142; table of contents in, 162; textual instruments, 150–51, **152**
Romney men, 71, 73
Romney method. *See* Dover Harbor, rebuilding of
Rotz, Jean, "Boke of Idrography," **104–5**
royal mining rights: grant of, 31, 33–34; leasing of, 52
Royal Navy: administration of, 3, 6, 7, 122–23; building of, 2–3, 97; Dover Harbor and, 55; merchant marine and, 4; need for copper, 36; Spanish Armada and, 2–6
royal patents: to Company of Mines Royal, 33–34; craft secrecy and, 15; joint-stock companies and, 9
Royal Society of London, 186–87
Russia trade, 89, 112, 116. *See also* Muscovy Company
rutter, 93–**95**; compiling of, 110, 116–17, 130–31; inaccuracy of data in, 94. *See also* navigational data

*Safeguard of Sailers, The* (Antoniszoon/Norman), **95**
Salomon's House. *See New Atlantis*
Sanderson, William, 88
*Scholemaster, The* (Ascham), 180
Schutz, Christopher, 38–39
Scot, Reginald: chronicle of Dover Harbor project, 55, 73, 226n. 1; on Dover commission, 60; on Francis Walsingham, 82; on John Trew, 63–64; Romney method and, 71
Scott, Thomas, 55, 68; overseer at Dover Harbor, 73–74, 84; Romney method and, 69–71
*Seamans Secrets, The* (Davis), 202; audience for, 163; black-boxing and, 164–65; format of, 163–64; latitude calculation, 164; textual instruments, 164
seawalls. *See* Dover Harbor, rebuilding of
secrecy: Baconian natural philosophy and, 205–6, 209, 211; copper battery and, 38–39; navigation and, 116–17; practical knowledge and, 15–17; Romney method and, 71
Sidney, Henry, 111
silver, 49; removal from copper ore, 28
Sixtus V (pope), 1
skill: as basis for expertise, 10–11, 213; definition of, 10; vs. experience, 11, 213
Smith, Thomas (the elder): funding for Dover Harbor repair, 83–84; leasing of royal mining rights, 52–53
Smith, Thomas (the younger), 241n. 32; mathematical lectureship and, 158, 161
Spain, population of, 1
Spanish Armada: defeat of, 5–6; preparations and departure of, 2, 5; Royal Navy and, 2–6

Spanish Empire, 1–2
Spanish-Habsburg army, 1–2
Spanish navigation, 3; exploration and, 98–99. *See also* navigation
Speydell, Sebastian, 32
spice markets of Asia, English quest for, 87, 91, 98
Sprat, Thomas, 186–87
staple markets, 88, 97
Stoultz, Leonard, murder of, 42–43, 224n. 55
surgery, 14
surveying, 138, 179–80; and German mines, **29–30**
Symons (mason from London), 74, 75

Taylor, E. G. R., 140–41, 183
technical treatises. *See* instructional manuals
Thames estuary, difficulty in piloting, 93
theoretical knowledge: as basis for expertise, 12, 16–17, 214; vs. empirical experience, 8; navigation and, 90; practical knowledge and, 15–16, 214
Thomson, John, 59, 79, 83
Thorne, Robert, 98
Thurland, Thomas, 39; difficulties as mine manager, 19–22, 33, 42–45, 46–49, 54, 85; lack of mining expertise, 20, 44–45, 54, 85, 215; petition to Queen Elizabeth, 32; role as expert mediator, 22, 35–36, 44–45, 54, 67, 215; royal patentee, 33–34
timber, English shortage of, 37, 40
Tirol region, German mining in, 31, 32
trade, overseas: decline of, 88, 97–98; drive to expand, 88–89, 97–98; English royal support for, 4, 97–98; Iberian monopolies, 98. *See also* Muscovy Company
"trained bands" of London. *See* civilian militia
traverse board, 93–94, **96**
Trew, John, 72, 215; career of, 228n. 18; dismissal of, 64; master surveyor at Dover Harbor, 63–66, 83

*Utopia* (More), 12–13

variation of the compass. *See* magnetic variation
vernacular: education, 14; instructional manuals, 140; translation of texts, 14, 220n. 28
*vita activa*. *See* active life
*vita contemplativa*. *See* contemplative life
Vives, Juan Luis, *De tradendis disciplinis*, 12
voyages of exploration. *See* exploration, voyages of

Wagenaer, Lucas Jansz, *The Mariners Mirrovr* (trans. Ashley), **100**
Walsingham, Francis: William Lord Cobham and, 63; William Courteney and, 4–5; Thomas Digges and, 55; Dover Harbor and, 72–73, 75–76, 78, 82; Richard Hakluyt and, 157–58; John Hawkins and, 77–78; Thomas Scott and, 55, 70, 74
Willoughby, Hugh: captain general for Muscovy Company, 112; death of, 87–88, 113; voyage of, 107, 112–13
wire pulling, 37–39. *See also* Company of Mineral and Battery Works
Worsop, Edward, defense of mathematics, 179–80
Wright, Edward: career of, 165; *Certaine Errors in Navigation*, 135, 165–69, 202, 203; lack of nautical experience, 165–66; mathematical lecturer, 165
Wynter, William: William Bourne and, 143; consultations on Dover Harbor, 61, 64, 69, 73

zinc mines: English, 4, 6, 39; German, 36